Non-Pharmacological Interventions

Gregory Ninot

Non-Pharmacological Interventions

An Essential Answer to Current Demographic, Health, and Environmental Transitions

 Springer

Gregory Ninot
University of Montpellier and Montpellier Cancer Institute
Montpellier, France

ISBN 978-3-030-60973-3   ISBN 978-3-030-60971-9   (eBook)
https://doi.org/10.1007/978-3-030-60971-9

This Springer imprint is published by the registered company Springer Nature Switzerland AG
The registered company address is: Gewerbestrasse 11, 6330 Cham, Switzerland

*To my friends Alain, Aline, Christophe,
David, Dominique, Estelle, François, Fred,
Greg, Isabelle, Jérôme, Jean-Marc, Jean-
Paul, Kim, Laurent, Laurel, Pierrick,
Raphaël, Simon, and Sophie
For Lou and Luka*

Preface

Nobody on our limited-resource planet can objectively ignore the huge challenges awaiting humans born in the twentieth century: climate change, aging of human populations, and unprecedented growth of noncommunicable diseases.

A vast list of solutions mixing traditional practices and modern technologies appears to be a fruitful avenue. Connected watches, serious games, health applications, virtual reality headsets, hypnoses, mindfulness, detox methods, diets, fasting, food supplements, enriched foods, minerals, herbal remedies, mycotherapy, essential oils, tisanes, massages, reiki, reflexology programs, osteopathy methods, acupuncture, spa treatments, Tai Chi, Qi Gong, Pilates, yoga, cryotherapies, light therapies, goop solutions...

Healthcare professionals, preventionists, consumers, patients, parents, family caregivers, and workers use and promote more and more these medicines, every day, everywhere. Media, social networks, and the Internet boost new consumers, workers seeking to be higher performing, athletes, students, retirees, patients wanting to obtain every chance of recovery, persons frustrated by conventional medicine, individuals wishing to get better in mind and body, and people looking for youth and attractiveness.

Where does this craze come from? Is it a profound and durable movement or a fashion with a momentary effect? Is this a credible health offer or a door open to the worst abuses? Is this sector to be taken seriously or does it remain marginal? Are there effective and safe practices? And if so, to cure, to care for, or to prevent which health troubles?

This book answers these questions through a scientific approach. Science and clinical research are revolutionizing this field (Ernst & Smith, 2018). In the course of 20 years, these practices have gone from global diet and hygiene strategies to targeted and personalized methods for prevention, care optimization, and even curative treatment.

Thanks to today's global research and its concern to identify effective and personalized solutions, dangerous practices are tending to be eradicated. From this immense shambles which used to mix alternative medicines, fake medicines, traditional remedies, useless interventions, and relevant solutions for health, science is now working to sort it out.

Patients will benefit from increasingly effective and safe health solutions, delivered at the right time, used by better and better-trained professionals. Scientists call them non-pharmacological interventions or NPIs. Their integration into a part of personalized medicine and integrative health is becoming increasingly evident (Rakel, 2018). It is clear that NPIs will play a major role in this century. They can be differentiated from alternative medicines (also called paranormal medicines) and socio-cultural services through continuous research that promotes a high-quality protocol and the traceability of practice. NPIs include approaches that go beyond promotional public health messages (e.g., generic hygienic-dietetic principles). Health authorities across governance levels are promoting improvements in their assessment and use (e.g., health agency, health ministry, medical society). Relevant NPIs should be proposed in tandem with regulated biomedical treatments. Unfortunately, not everyone is benefitting from NPIs at this time, as only the richest of the rich countries have the knowledge, the time, and the means to use them wisely. As such, for the moment, only the poorest of the developing countries that lack the biotechnological means to cure their populations are inspired by NPIs as they draw from their own traditional medicines.

NPIs have become essential solutions for better living, for preventing diseases, for treating diseases in addition to biomedical therapies, and for increasing longevity without the loss of quality of life. They include practices that have both been selected empirically for centuries and that have appeared recently with the help of technological revolutions, epigenetics, clinical research, and socioeconomics. As a result, NPIs are increasingly developing and diversifying around the world. A real ecosystem of a myriad of public and private actors is in construction.

Like any new disruptive approach, NPIs must take their place – their full place – nothing but their place. They should not be used to replace biomedical therapies, but rather to complement them and to satisfy the needs of quality of life and healthy aging. The ethical practice of NPIs should not provide foolproof guarantees or give rise to any form of charlatanism or abuse. As such, the standardization and labeling of these practices and their professionals is becoming increasingly essential, as well as increasingly urgent.

The book analyzes this phenomenon on a growing scale, destined to overturn lasting prevention and care practices. It offers keys to understand and use them wisely.

Montpellier, France   Gregory Ninot

References

Ernst, E., & Smith, K. (2018). *More harm than good? The moral maze of complementary and alternative medicine.* Cham: Springer.

McIntyre, L. (2019). *The scientific attitude. Defending science from denial, fraud, and pseudoscience*. Cambridge: Massachusetts Institute of Technology.
Rakel, D. (2018). *Integrative medicine* (4th ed.). Philadelphia: Elsevier.

Acknowledgments

This book is first and foremost a mark of gratitude to all the patients and caregivers who have shown me the path to NPIs.

I also thank colleagues, collaborators, students, and anonymous reviewers who shared this quest for a rigorous scientific approach in a field so sensitive to bias and beliefs.

This book pays homage to my mentors and sources of inspiration: Christophe André, Jean Bilard, Jean Bourbeau, Gérard Bourrel, Jean Bousquet, Jacques Bringer, Jacques Desplan, Bruno Falissard, John Ioannidis, Jacques Kopferschmitt, Julien Nizard, François Paille, Lynda Powell, Christian Préfaut, Henri Pujol, Lise Rochaix, Pierre Senesse, Daniel Serin, and Daniel Schwartz.

This book shows my gratitude to the French university system, and more particularly to that of Montpellier, which hosts the oldest medical school in Europe.

Contents

About the Author

Gregory Ninot PhD, is full professor at the University of Montpellier and at the Montpellier Cancer Institute in France. He has followed a dual curriculum in sports sciences (master's degree, PhD) and psychology (master's degree, PhD) applied to health. Since 1999, his basic research has focused on integrated models explaining psychological and behavioral adjustment to chronic disease and healthy aging. His applied research focuses on assessing the benefits, risks, and cost-effectiveness of non-pharmacological interventions (NPIs). Professor Ninot is the author of more than 150 articles in international scientific and medical journals and 11 books. He has led several multidisciplinary academic research units interested in health issues. Professor Ninot founded and currently manages a university collaborative platform, the Plateforme CEPS; organizes an annual international conference, the iCEPS Conference (www.icepsconference.fr); and develops digital tools (e.g., www.motrial.fr, www.nishare.fr, www.niri.fr) to improve NPI assessment and evidence-based practices. Since 2014, he has run a blog on NPIs (www.blogensante.fr/en/).

Abbreviations

ACT	Acceptance and Commitment Therapy
ANSM	National Agency for Drug Security (France)
CAM	Complementary and Alternative Medicine
CBT	Cognitive Behavioral Therapy
COPD	Chronic Obstructive Pulmonary Disease
CPMDQ	Complementary Medicine Professional Corporation of Quebec (Canada)
CRIOC	Consumer Organization Research and Information Center (Belgium)
CRO	Contract Research Organization
EBM	Evidence Based Medicine
EFSA	European Food Safety Authority (Europe)
EFT	Emotional Freedom Technique
EMA	European Medicines Agency (European)
EMDR	Eye-Movement Desensitization and Reprocessing
FDA	Food and Drug Administration (USA)
GIW	Global Institute of Wellness
HAS	High Health Authority (France)
HRQL	Health Related Quality of Life
ICTRP	International Clinical Trials Registry Platform
INCA	National Cancer Institute (France)
INSERM	National Medical and Health Research Institute (France)
MBSR	Mindfulness Based Stress Reduction
NCCIH	National Center for Complementary and Integrative Health (USA)
NICE	National Institute for Care and Health Excellence (Great-Britain)
NIH	National Institutes of Health (USA)
NPI	Non-Pharmacological Intervention
OECD	Organization for Economic Co-operation and Development
RCT	Randomized Controlled Trial
WHO	World Health Organization

Chapter 1
Defining Non-pharmacological Interventions (NPIs)

1 Introduction

Alternative medicines, complementary medicines, natural medicines, traditional medicines, integrative medicines, preventive medicines, lifestyle medicines, health-care solutions, antiaging protocols, goop methods, grandmother remedies, etc., are part of our daily lives. They seek to heal illness, to delay the effects of aging, to prevent health conditions, to improve wellness, to enhance work performances, in other words, to enhance well-being, autonomy, and longevity. Perhaps you have already tried reflexology, chiropractic, shiatsu, herbal detox teas, fasting, meditation, and smartphone health applications? How can we clarify these remedies with a scientific approach?

2 Inventory

2.1 Practices, Techniques, Remedies, and Methods

Do you remember what your drugstore was like in the last century, 20 years ago? A pharmacist who had graduated from a long scholarship at a university welcomed you from behind her or his counter. She or he read the doctor's prescription, checked all the details, and gave you the relevant, prescribed medications with advice on how to take them. She or he politely answered all of your questions, even the stupidest and most shameful. Nowadays, the counter no longer exists. The pharmacist is becoming increasingly rare, replaced by a commercial agent. The border between the pharmacy and the drugstore is blurring. Knowing how to navigate through all the supposedly healthy stalls is a challenge. The pharmacy is full of shelves of vitamin cocktails, detox mineral powders, plant extracts for sleep, anti-cholesterol capsules made from mushrooms, antioxidant dried seaweed, anti-bloating coconut charcoal, fortifying tonics of trace elements, toning tablets, antiaging capsules, fat burners, anti-cellulite creams, bottles of anti-stress essential oils, and soothing herbal teas. Pharmacies are becoming increasingly integrated in commercial centers, blurring the frontier between medical/health and consumer products.

Similarly in bookstores, shelves devoted to health and personal development abound. Everyone goes with their own method, their essential solution, their original advice, their long experience. For example, celebrity doctors – like Ginni Mansbergin in Australia; Eriko Wakisakain in Japan; Michel Cymes and Frederic Saldmann in France; Pixie McKenna, Dawn Harper and Christian Jessen in the United Kingdom; Deepak Chopra, Mehmet Oz, Drew Pinsky, David Perlmutter, and Juan Riviera in the United States; Nicola Romero and Miguel Angel Martínez Gonzalez in Spain – all offer packages of non-pharmacological solutions for better health in their respective books, which have become bestsellers translated into several languages. Medical professors like David Khayat and Henri Joyeux give advice, some of which hit the headlines. Some became famous only for a specific diet like Robert Atkins, Arthur Agatston, Jean-Michel Cohen, and Pierre Dukan. Well-known figures from prestigious medical schools explain that the U.S. Department of Agriculture (USDA) guidelines are not only wrong, but also dangerous and recommend 77 "new and revolutionary remedies to be healthy" (Willett and Skerrett 2011). Famous clinics "offer their innovative home remedies with attractive advertisement (see box)." Intellectuals like Jill Bolte Taylor in the United States or Michel Onfray in France testify for natural solutions that they claim saved their lives after a stroke. Researchers like Nicole Apelian and Claude Davis explore lost herbal remedies that kept previous generations alive and share "550+ powerful natural remedies made from them for every one of your daily needs" (Apelian and Davis 2018). Movie stars like Cameron Diaz, Jane Fonda, and Gwyneth Paltrow; singers, athletes, models, and restaurant chiefs; all of them share their recipes for better health and longevity.

The Mayo Clinic Book of Home Remedies Published in 2017
"Learn preventative measures to minimize the risks of many everyday health issues. Then, improve your ability to diagnose just what might be wrong by considering the signs and symptoms, causes, and possible outcomes of more than 60 common ailments, from the stomach flu and back pain to depression or diabetes. If conditions do develop, Mayo Clinic Home Remedies presents practical steps you can take at home before going to the doctor. Such effective self-care can save you money and the hassle of dealing with doctor's appointments. But you might also feel a new sense of autonomy and control in the chaotic world of personal health care. In the event of more a more serious condition, additional Medical Help advises you on when to seek medical attention. Alphabetically arranged for quick reference, this comprehensive and authoritative guide will help you stay healthy, feel energized, independent, and enjoy a higher quality of life." Efficient and safe recommendations? Perhaps...

Psychotherapies are multiplying – and when possible are avoiding mentioning the word "therapy" while indicating their trademark (®): eye-movement desensitization and reprocessing (EMDR), neurolinguistic programming, sophrology, systemic constellations and systemic family, Tipi emotional regulation, and the Tomatis Method. Some are inspired by oriental meditation such as mindfulness-based stress reduction (MBSR), Vipassana meditation, the emotional freedom technique (EFT), and reconnective healing. Hypnosis is gaining more and more users and is being applied to everything: quitting smoking, reducing pain, and treating phobias. Some mind-body practices, such as restorative yoga, anti-gravity yoga, Siddha yoga, Qi Gong, or cardiac coherence protocols are fashionable. Professionals promise you better health, lasting happiness, less stress, and better resistance to difficult life events, microbes, viruses, etc.

With technological and miniaturization progress, tools are now available to monitor your everyday health. For example, a weight scale now shows your body composition (muscle, bone, water, and fat mass) in addition to your weight and gives you personalized advice. Connected to your mobile phone, it tells you what to eat. It predicts your weight change and it alerts you if your personal minimum and maximum thresholds are exceeded. Other objects monitor your biological constants such as blood pressure, blood sugar, or heart rate and evaluate behavioral parameters such as daily step-counts or time spent in a sitting position. These wearable objects are diversifying: bracelets, watches, armbands, badges, belts, soles, shoes, clothes, patches, rings, necklaces, tactile nails, glasses, and electronic cigarettes. Everyday devices love your health data: fridges, ovens, food processors, coffee makers, glasses, spoons, straws, food composition analyzers (e.g., gluten detector), thermometers, sleep regulators, alarm clocks, mirrors, toothbrushes, electronic pill organizers, robots, digital pets, children's toys, electronic games, and video games. Others follow you in your leisure and your travels: on-board mapping systems, car driving analyzers, bicycles, gyro-wheels, and rackets. Others are installed at work: keyboards, armchairs, vending machines, cameras, helmets, glasses, and styluses. All the fans rush every mid-January to the Las Vegas CES meeting to discover the latest innovations of engineers who out-nerd those of the previous year. The conference draws more than 170,000 attendees from 160 countries. As if that were not enough, we voluntarily or unknowingly post personal information about our health via social media; if only by making selfies or videos visible – a photo compared to others says a lot about our health in and of itself. Additionally, our favorite Internet search engine knows our tastes and preferences. Without even knowing it, we regularly share direct and indirect medical data on individual and/or collective messaging. Last, but not least, remains our dear, supposedly intelligent pocket or bag companion: the smartphone. Seventy-seven percent of people are equipped with this tool, which has become as vital as it is viral. It follows us everywhere and interrupts our conversations with its untimely sounds and subtle notifications. It measures, photographs, memorizes, integrates, analyzes and shares personal behavioral, biological, economic, and social data transmitted more or less with our agreement – the quantified self. It plans our life. Today it helps us make decisions, and thanks to the spread of artificial intelligence, tomorrow it will make them for us. Beyond its initial function for communication, this intelligent object – with our personal data sorted, cross-checked, consolidated, and compared to huge databases – becomes a virtual health advisor. Do you still think that these connected objects are only gadgets or that the data they collect is useless?

Information technologies change in the span of a few decades. Media, the Internet, and social media all expand information. Real experts, so-called experts, who repeat what they heard on some form of media a few hours ago, and fake experts, who launch fake news hoping to make a buzz, say everything and its opposite. The COVID-19 pandemic reminds us that the more experts there are, the less we understand. Since health is one of the three areas of interest for smartphone users – along with food and sports – one in two smartphone owners have downloaded at least 1 of the 318,000 health apps available, not to mention their consultation of health websites. For example, among all the available information, you had probably discovered original practices such as *intermittent fasting* or *sylvotherapy*. You have not resisted transmitting to your friends and to social networks in order to have fun, to be original. One day, a more vulnerable, a more sensitive, or a more curious person will succumb to it. And so, that is it — the "virus" is transmitted. Books, videos, audio recordings, manuals, internships, trainings, certifications, the practice is expanding irremediably around the world.

Example of Sylvotherapy

Sylvotherapy, forest bathing or *Shinrin-Yoku*, is a Japanese practice for treating stress using techniques for connecting to trees, and more particularly to a tree. The *sylvotherapy* professional displays, in all seriousness, her or his "official certification and qualification" from the Japanese master, Qing Li, who claims to have published a study in a scientific journal (Li 2010). Despite the claims found on the web by interested followers, no study has shown any health benefit from this practice. A walk in the forest cannot hurt, but going from there to making it a stress therapy shows us how easy it is to make scientific and medical shortcuts that make people believe in the effectiveness of a method. It is enough to juxtapose a health problem, a human practice, a recipe (inspired by traditional techniques or created ex nihilo), an action mechanism, a scientific reference, hand-picked testimonies, and a few passages in media that is eager for novelties, and the turn is played. The worst part is that the people who engage in these practices are not necessarily malicious. Simply put, their personal beliefs and experiences have taken precedence over science. Their lack of scientific training and predisposition to believe (ignoring cognitive biases) leads them toward fanciful arguments. If medicine is an art, do you think it should be practiced by artists?

Twenty years ago, massages were rarely proposed as a therapy in the Western world with the exception of physiotherapists following targeted medical prescription for musculoskeletal pain, severe limitations, and elite athletes. Now, you can

Table 1.1 100 examples of massage protocol

1. Abhyanga massage	51. Metamorphic massage
2. Access bars	52. Moneyron method
3. Acustimulation	53. Myofascial massage
4. Akumat therapy	54. Myofascial release
5. Amma-seated massage	55. Neo-Reichian massage
6. Anthroposophical rhythmic massage	56. Neuro-reflexotherapy
7. Aromatherapy massage	57. Niromathe method
8. Auriculoreflexology	58. NTE massage
9. Auriculotherapy	59. Pressure therapy
10. Ayurveda massage	60. Psycho-bio-acupressure
11. Balinese massage	61. Radloff massage
12. Baunscheidt method	62. Rebalancing
13. Biodynamic massage	63. Reflex area massage
14. Biosynthesis massage	64. Reflexology
15. Body base manipulative therapy	65. Reflexo-therapy
16. Body Poyet method	66. Reiki therapy
17. Bowen technique	67. Release of the pericardium
18. Californian massage	68. Remedial massage
19. Chi Nei Tsang	69. Rhinoreflexology
20. Colon massage	70. Rhythmic massage
21. Conjunctive tissue massage	71. Rolfing massage
22. Craniosacral therapy	72. Rolfing structural integration
23. Danis bois method	73. Scar mobilization
24. Deep tissue massage	74. Sensitive gestalt massage
25. Empirical massage	75. Sensitive massage
26. Esalien massage	76. Sensitive wellness massage
27. Facial morph intervention	77. Shin tai or shiatsu massage
28. Fasciapulsology	78. Skin heal
29. Fasciatherapy	79. Skin needle massage
30. Foot reflexology	80. Somatopathy
31. Gentle massage	81. Sonopuncture
32. Gentle touch	82. Spinal manipulative therapy
33. Gesret method	83. Spine base balancing therapy
34. Hakim massage	84. Sumathu massage
35. Hand massage program	85. Swedish massage
36. Hand reflexology	86. Tantra massage
37. Handling therapy	87. Tapas acupressure technique
38. Hawaiian massage	88. Technical acupressure tapas
39. Healing touch	89. Thailand or Thai massage
40. Hellerwork structural integration	90. Therapeutic massage
41. Intuitive wellness massage	91. Trager approach
42. Jin shin do	92. Trigger point therapy
43. Kashmiri massage	93. Tui Na or Tuina massage

(continued)

Table 1.1 (continued)

44. Korean manual puncture	94. Vertebrotherapy
45. Lemniscate massage	95. Vitalogy massage
46. Lomi-Lomi massage	96. Vodder method
47. Lymphatic drainage massage	97. Water touch art
48. Manual medicine technique	98. Well-being massage
49. Psychogenic areas massage	99. Zero balancing
50. Metamorphic massage	100. Zusanli

benefit from a "therapeutic" massage everywhere – airport terminals, bars, hotels, and street corners – for everything. No limits. Do you know or recognize your last massage among the list below? (Table 1.1)

2.2   Professionals

Faced with high-tech medicine, with a depersonalized, sanitized, and sterilized hospital, with approaches that aim to treat a deficient organ rather than a human being living in a given environment and context, traditional medicine and its professionals captivate. How can you not be attracted to someone who really takes time to listen to you, who takes a real interest in you, who gives you explanations of the evils you suffer from, and who gives you hope and confidence in your self-power to recover? What's more, all the better if the first session is free of charge!

So many professions related to heath, well-being, wellness, and fitness have been created in the last 10 years. Professional plaques appear on the walls of buildings with names that seem as obscure as the methods and training of those who practice them. Everywhere you go, they claim to have a holistic view of the human. Practitioners differ from medical workers, such as general doctors, who spend an average of 12 minutes per patient, much of which is spent completing administrative requests on her or his computer. These professionals can have exotic names without extending to spiritual, sorcery, marabou, or prophetic fields.

> **An Unlimited List of Professionals**
> Anthroposopher, art therapist, body-mind practitioner, bodyminder, bone repairer, dowser, endobiogenist energetician, fire cutter, geobiologist, healer, health coach, magnetizer, medicine man, micro-nutritionist, micro-physiotherapist, music-therapist, naturopath, naturopathic doctor, nutripuncturist, pet-therapist, psychosomatician, rebirthing practitioners, traditional healer, shaman, witchdoctor, zootherapist.

Some professionals have difficulties clearly explaining their practice, hiding behind esoteric arguments, strange formulations, and unknown abbreviations. How can we identify these trades and ensure the skills of the people who practice them?

Faced with this surge of health care offers, everyone seems to be left to their own views. A benefit, even an infinitesimal one, is automatically attributed to the practitioner. A psychological insight into the intimate life of the client supports that trust. A bodily sensation supports a testament to pain liberation. All of the client's attention is focused on finding confirmation of the expected effect, and at the same time, to avoiding feeling like an idiot with no return on investment. A positive experience gives the right and the ability to recommend the professional using the famous formula: "at least, it can't hurt you." Everyone gives their opinion on social networks and accelerates the dissemination of these recipes that are said to be favorable to human health, but which are most often based on individual satisfaction. Singular experiences become collective truths. Virtual communities are created across frontiers and beyond regulations.

2.3 Organizations

Wellness centers, antiaging camps, integrative health hospitals, integrative medicine institutes, and detox establishments flourish on all continents. To be attractive, these institutions use vocabulary with fashionable hybridity, such as CrossFit or Aquabiking; with Indian inspiration, such as Ayurveda or Unani; or with Asiatic influence, such as Qi and its declinations. These places are soothing, the atmospheres hushed, and the staff friendly. Treatments are "à la carte": from a day to a few days in a "detox," or "natural health" courses in unusual places, and not forgetting the manual therapies offered in stations and airports.

Trade fairs on well-being, hydrotherapy, naturopathy, natural health, and sport-health are multiplying the territory. They advertise themselves on billboards, on television, in newspapers, and even on specialized blogs.

All non-pharmacological practices have been extensively developed and diversified in these institutes since the beginning of the twentieth century. They respond to contemporary needs of weak humans living in violent, inequitable, uncertain, technically dependent, and individualistic societies in a chaotic and unpredictable environment. They are no longer the effect of a passing fashion. They are now a lasting part of our lives. Alas, the best meets the worst; a fitness trail can quickly become an obstacle course. It's time to sort the good from the chaff! It is also time to appropriately qualify these methods.

3 Authorities' Difficulties

3.1 Approach Level

How to name non-pharmacological practices and limit their perimeter?

The World Health Organization (WHO) (2013) proposes the concept of *"traditional and complementary medicine"* (T&CM). Without precisely defining its boundary, this term describes techniques, therapeutic methods and approaches, and also medical systems. In a recent document, *"the terms 'complementary medicine' and 'alternative medicine' refer to a broad set of health care practices that are not part of that country's own traditional or conventional medicine and are not fully integrated into the dominant health care system. They are used interchangeably with traditional medicine in some countries"* (WHO 2019a, b, p.8). It is, therefore, a vast area where diagnostic, preventive, therapeutic, educational, and lifestyle approaches can be mixed.

> **Traditional Medicine According to WHO (2020)**
> *"Traditional medicine has a long history. It is the sum total of the knowledge, skill, and practices based on the theories, beliefs, and experiences indigenous to different cultures, whether explicable or not, used in the maintenance of health as well as in the prevention, diagnosis, improvement or treatment of physical and mental illness."*

> **Complementary Medicine According to WHO (2020)**
> *"The terms "complementary medicine" or "alternative medicine" refer to a broad set of health care practices that are not part of that country's own tradition or conventional medicine and are not fully integrated into the dominant health-care system. They are used interchangeably with traditional medicine in some countries."*

In Australia, the main term used is *"complementary medicines"* following a 2011 report of the Australian National Audit Office about therapeutic goods regulation. *"Medicinal products containing herbs, vitamins, minerals, and nutritional supplements, homoeopathic medicines and certain aromatherapy products are referred to as 'complementary medicines'. Complementary medicines comprise traditional medicines, including traditional Chinese medicines, Ayurvedic medicines and Australian Indigenous medicines. Other terms sometimes used to describe complementary medicines include 'alternative medicines', 'natural medicines' and 'holistic medicines'. Complementary medicines are generally available for use in self-medication by consumers and can be obtained from retail outlets such as phar-*

macies, supermarkets and health food stores. While the majority of complementary medicines are indicated for the relief of symptoms of minor, self-limiting conditions, many are indicated for maintaining health and wellbeing, or the promotion or enhancement of health (p. 33)". In fact, the official definition of the Therapeutic Goods Administration (2020) restricts *"complementary medicines"* to therapeutic goods consisting principally of one or more designated active ingredients:

- An amino acid
- Charcoal
- A choline salt
- An essential oil
- Plant or herbal material, including plant fibers, enzymes, algae, fungi, cellulose, and derivatives of cellulose and chlorophyll
- A homeopathic preparation
- A microorganism, whole or extracted, except a vaccine
- A mineral including a mineral salt and a naturally occurring mineral
- A mucopolysaccharide
- Nonhuman animal material including dried material, bone and cartilage, fats and oils, and other extracts or concentrates
- A lipid, including an essential fatty acid or phospholipid
- A substance produced by or obtained from bees, including royal jelly, bee pollen, and propolis
- A sugar, polysaccharide or carbohydrate
- A vitamin or provitamin

Australian researchers are more inclusive. The Western Sydney University's NICM Health Research Institute (NICM) divides *"integrative healthcare approaches"* into two subgroups: natural products (herbs, vitamins and minerals, and probiotics) and mind and body practices (yoga, chiropractic and osteopathic manipulation, meditation, massage therapy, acupuncture, relaxation techniques, tai chi, qi gong, healing touch, hypnotherapy, and movement therapies). The Australian Research Centre for Complementary and Integrative Medicine (ARCCIM) proposed *"complementary and integrative medicine"* (CIM) to include a wide range of healthcare practices and therapies not currently associated with the medical profession or the medical curriculum, such as acupuncture, naturopathy, herbal medicines, chiropractic, and massage among others.

In China, the law on *"traditional Chinese medicine"* (TCM) was approved by the National People's Congress in 2016.

In France, the health ministry uses the generic term *"unconventional care practices"* without categorizing them. Several official reports mention that "these practices are diverse" and suggest indifferently employing *"complementary medicine," "natural medicine,"* or even *"alternative medicine."* In 2011, the French National Authority for Health (HAS) recommended the term *"non-pharmacological therapies."*

In Great Britain, the National Health Service (NHS 2020) uses the term "*complementary and alternative medicines*" (CAM). These CAMs correspond to treatments that are not part of conventional health care.

In India, Ayurveda, yoga and naturopathy, Unani medicine, Siddha and Sowa Rigpa, and homoeopathy for holistic health care were combined in a new holistic approach called AYUSH. Boards/councils for the registration of practitioners, hospital departments, and dispensaries were created in order to facilitate the integration of AYUSH into the health services delivery network.

> **AYUSH**
> In India, Ayurveda, yoga and naturopathy, Unani medicine, Siddha and Sowa Rigpa, homoeopathy for holistic health care were combined in a new integrative approach called AYUSH, combined with conventional medicine.

In Iran, "*Persian medicine*" is one of the oldest traditional systems of medicine dating back to at least 7000 years ago. It flourished during the Islamic Era, in particular from the ninth to the thirteenth century, and it was the main paradigm of medicine in most parts of West and Central Asia, Africa, and Europe until the seventeenth century. It has a strong philosophy and generation-by-generation experiences that make it a good source for health care and treatment ideas. Persian medicine is accepted by the people as it matches their culture and traditional beliefs.

In Morocco, complementary medicine is mainly represented by traditional medicine. This "*traditional Moroccan medicine*" has its origins in Arab and Islamic medicine. It occupies a preponderant place in the life of the population, corresponding to the total sum of knowledge, skills, and practices that are based, rationally or not, on the theories, beliefs, and experiences specific to Moroccan culture. This medicine is mainly based on the use of medicinal and aromatic plants.

In Norway, the National Research Center in Complementary and Alternative Medicine (NAFKAM) uses the term, "*complementary and alternative medicine.*"

In Turkey, the government uses the term, "*traditional and complementary medicine practice*" after a regulation passed in 2014.

In the United States, Congress passed legislation in 1992 that provided $2 M USD in funding for the establishment of an office within the National Institutes of Health (NIH) to investigate and evaluate promising "*unconventional medical practices.*" In 1999, the National Center for Complementary and Alternative Medicine (NCCAM) popularized "*complementary and alternative medicine*" accords around the world. The term CAM was consolidated within the national academies' Institute of Medicine report published in 2005. In 2014, Congress renamed NCCAM as the National Center for Complementary and Integrative Health (NCCIH). The terms "*complementary and integrative medicines,*" "*complementary and integrative therapies,*" and "*complementary health approaches*" were used. While the term integrative refers to using a holistic approach to the person (considering the whole patient), nothing specifies the ranking of solutions or their use as a function of time. The term integrative mixes health problem assessments and solutions. In addition, while in the singular the concept of integrative medicine can make sense (the association of methods to maximize the chances of curing or preventing disease), in the plural it can become contradictory or even absurd (several juxtaposed approaches do not potentiate or are opposing).

Challenge of a Single Definition
"The use of the terms 'complementary' and 'alternative' serve only to detract from a therapy by making it sound second class. Therapies that are often labeled under the heading of CAM include nutrition and spirituality. Many would argue that a lack of attention to these important influences on health has resulted in epidemics of obesity, diabetes, and substance abuse." (Rakel and Weil 2018, p. 4).

In Saudi Arabia, *"complementary therapies or approaches"* is included in the integrative model of healthcare transformation (Khalil et al. 2018). This model integrates both well-researched and culturally appropriate practices. The advantage of complementary medical intervention is its holistic approach, which is best suited to patients who have particular psychological and spiritual needs. Holistic approaches pay attention to these needs, focus on the patient-doctor relationship, and understand the patient's perspective through multimodal concepts. The rationale behind integrative medicine, then, is to include the best practices of both conventional and complementary therapy, uniting these practices into a holistic approach that offers the benefits of both approaches.

Challenge of a Single Definition
"A worldwide definition of CAM looks like far from being realistic. The first obstacle is the different views between East and West" (Casarin et al. 2019, p.2).

Depending on a Western or Asian approach, the country, the health profession, and the preventive or therapeutic viewpoint, practices can be named in English by 100 terms that are similar, but not exactly synonyms (Table 1.2). No one term predominates in either the world or in an individual country. In certain medical specialties, they are called supportive care, and even pejoratively as comfort care, cuddle therapies, or placebos.

3.2 Method Level

As soon as you get into the details and down to the level of method (protocol, program, preparation, and procedure), it feels as if you are opening a Pandora's box where everything becomes elusive. How many methods? For what kinds of health problems? Furthermore, many authors of questionable reviews are clueless. They often conclude their article, report, or book with a cryptic sentence expressing regret

Table 1.2 100 appellations far from being synonymous

1. Adjunctive interventions	51. Natural remedies
2. Adjuvant care	52. Noninvasive approaches
3. Adjuvant therapies	53. Nonconventional medicines
4. Alternative health therapies	54. Nonconventional care
5. Alternative medicines	55. Non-pharmacological approaches [a]
6. Alternative remedies	56. Non-pharmacological interventions [a]
7. Anthroposophical medicines	57. Non-pharmacological managements [a]
8. Behavioral medicines	58. Non-pharmacological measures (WHO 2014) [a]
9. Body-mind methods	59. Non-pharmacological methods [a]
10. Body and mind therapies	60. Non-pharmacological practices [a]
11. Care interventions or protocols	61. Non-pharmacological processes[a]
12. Comfort care	62. Non-pharmacological protocols[a]
13. Complementary and alternative medicine (NCCAM 1999)	63. Non-pharmacological recipes[a]
14. Complementary health approaches (NCCIH 2016)	64. Non-pharmacological remedies[a]
15. Complementary and integrative alternative medicines	65. Non-pharmacological solutions[a]
16. Complementary medicines (WHO 2018)	66. Non-pharmacological strategies[a]
17. Complementary nursing practices	67. Non-pharmacological therapies[a]
18. Complementary practices	68. Non-pharmacological treatments[a]
19. Complementary therapies (French Academy of Medicine 2013)	69. Non-pharmacotherapies
20. Complementary treatments	70. Nondrug interventions
21. Complex or multimodal interventions	71. Nonpharmaceuticals
22. Comprehensive health approaches	72. Occupational health interventions
23. Comprehensive managements	73. Palliative interventions
24. Cuddle therapies	74. Parallel medicines
25. Energetic care	75. Paramedical care
26. Evidence-based interventions in prevention and health promotion	76. Paramedical protocols or techniques
27. Grandmother or home remedies	77. Paramedical treatments
28. Healer care	78. Person-centered care
29. Health claims	79. Personal development services
30. Health interventions (WHO 2017a, b)	80. Placebos
31. Health practices	81. Personalized complementary therapies
32. Health prevention actions	82. Positive health practices
33. Holistic care	83. Preventive medicines
34. Holistic medicines	84. Psychosocial interventions
35. Hygiene and dietetic interventions	85. Psychosocial rehabilitations
36. Integrated care techniques	86. Psychosocial rehabilitation care
37. Integrative and comprehensive health interventions	87. Psychosomatic therapies

(continued)

Table 1.2 (continued)

38. Integrated care	88. Public health interventions (WHO 2017a, b)
39. Integrative health solutions	89. Rehabilitation methods
40. Integrative healthcare approaches (NICM 2020)	90. Replacement therapies
41. Integrative medicines	91. Routine care protocols
42. Lifestyle medicines	92. Self-care remedies
43. Mind and body practices	93. Support interventions
44. Multimodal therapy	94. Supportive care
45. Natural complexes	95. Technological aids
46. Natural cure	96. Traditional medicines
47. Natural healing remedies	97. Traditional and complementary medicines
48. Natural health preparations	98. Unconventional care practices (French Ministry of Health 2017)
49. Natural medicines	99. Validated non-pharmacological therapies
50. Natural products	100. Wellness care techniques

[a]non-pharmacological or nonpharmacological, non-pharmacological or non-pharmaceutical, non-pharmacologic or nonpharmacologic, non-pharmacological or non-drug

over the heterogeneity of practices that prevents them from providing a clear answer to the question they pose at the outset: is this a solution that is effective and safe for health?

For example, there are over 200 essential oils. They can be administered via dermal, respiratory, or internal routes. They can have analgesic (pain, skin irritations, insect bites, blows, hematoma), anti-stress, antiseptic, digestive, or respiratory properties. They carry risks, including the potential to cause irritation, allergies, or sun reactions. They are not recommended for children under 3 years of age, pregnant or breastfeeding women, or for people with serious health problems. This aromatherapy example provides insight into the multitude of methodological questions surrounding one practice, while in 2013 the French Senate listed 115 practices in 2013 (Table 1.3). It can be found in:

– University disciplines (plant science or psychology, for example)
– Paramedical disciplines (physiotherapy, for example)
– Health professions (liberal caregiver, for example)
– Methods (EMDR, for example)
– Techniques (tapas acupressure skill, for example)
– Diagnostic approaches (iridology, for example)
– Lifestyles (Feng Shui, for example)
– Traditional medicines (Chinese medicine, for example)
– Condemned sectarian practices

A Belgian institute identified 12 more practices not included in the French Senate list (*Centre de Recherche et d'Information des Organisations de Consommateurs* 2012) presented in Table 1.4. This type of census, based on interviews and Internet

Table 1.3 Alternative medicine list according to the French Senate. (French Senate 2013)

1. Acupressure	59. Abhyanga massage
2. Acupuncture	60. Amma-Assis massage
3. Analysis and re-information	61. Californian massage
4. Aromatherapy	62. Hakim massage
5. Aromatology	63. Intuitive Well-being massage
6. Art-therapy	64. Leinmscate massage
7. Atlas-profilax	65. Sensitive massage
8. Auratherapy	66. Tui Na massage
9. Auriculoreflexology	67. Ayurveda medicine
10. Auriculotherapy	68. Hildegarde de Bingen medicine
11. Baubiology	69. Traditional Chinese medicine
12. Bioenergy	70. Gesret method
13. Biomagnetism	71. Mezieres method
14. Bioresonance	72. NAET method
15. Holistic biotherapy	73. Vittoz method
16. Bowen NST	74. Milta-therapy
17. Chiropractic	75. Musicotherapy
18. Chromotherapy	76. Naturopathy
19. Life coaching	77. Nutritherapy
20. Bach flower consulting	78. Olfactotherapy
21. Family and systemic constellations	79. Oligotherapy
22. Dance therapy	80. Ondobiology
23. Biological decoding	81. Osteopathy
24. Holistic dentistry	82. Ozonotherapy
25. Dien Chan	83. Phytotherapy
26. Chinese dietetics	84. PNL
27. Digitopuncture	85. Psycho-bio-acupressure
28. Do in	86. Psycho-energy
29. Manual lymphatic drainage	87. Psycho-genealogy
30. Poyet listening to the body	88. Psychonomy
31. EFT	89. Psychopractice
32. Electropuncture	90. Psychotherapy
34. EMDR	91. Qi gong
35. Enneagram	92. Quantum touch
36. Energy balance	93. Radiesthesy
37. Etiopathy	94. Rebirth
38. Fasciapulsology	95. Reboutement
39. Fasciatherapy	96. Reflexology
38. Feng Shui	97. Reiki
40. Focusing	98. Relaxation
41. Geobiology	99. Holotropic breathing
42. Gestalt therapy	100. Liberal midwife
43. Healer	101. Shiatsu

(continued)

Table 1.3 (continued)

44. Ehrenfried holistic gymnastic	102. Somatopathy
45. Hirudotherapy	103. Somatotherapy
46. Homeopathy	104. Sophrology
47. Hydrotherapy	105. Tapas acupressure technique
48. Hypnosis	106. Alexander technique
49. Iridology	107. Energy harmonization technique
50. Irrigation of colon	108. Nadeau technique
51. Kinesiology	109. Craniosacral therapy
52. Kinesitherapy	110. Therapy of mental field
53. Reconnexion	111. Sound therapy
54. Trame	112. Tinnitometry
55. Lithotherapy	113. Tipi
56. Magnetism	114. Yoga
57. Magnetotherapy	115. Zensight
58. Korean manupuncture	

Table 1.4 Supplementary practices identified by a Belgian institute. (Centre de Recherche et d'Information des Organisations de Consommateurs 2012)

116. Immaterial biosurgery	122. Morphopsychology
117. Total biology	123. Neoshamanism
118. Shamanism	124. Pelotherapy
119. Meditation	125. Phototherapy
120. Feldenkrais method	126. Sympathicotherapy
121. Pilates method	127. Tai chi

searches, reveals both a great heterogeneity of practices and a probable non-exhaustiveness. Not all areas are covered by this kind of exploratory approach; for example, we could add: diets, food supplements, plants remedies, mineral extracts, mushroom fragments, adapted physical activity programs, health-connected objects, ergonomic objects, physiotherapy methods, psychotherapy protocols... Without a systematic, continuous and methodic approach, it will be impossible to identify and hierarchize these practices.

The French National Authority for Health (HAS) distinguishes three categories of non-pharmacological therapies: "physical therapies," "hygienic-dietetic rules," and "psychological treatments," without going so far as to list all the methods.

Norway's NAFKAM categorized the use in terms of "going to a provider," "herbs/natural remedies," and "self-help techniques."

The U.S. NCCIH distinguishes these three categories without detailing them: natural products (herbs, minerals, vitamins, probiotics, etc.); mind-body practices (e.g., yoga, chiropractic, osteopathy, massage, meditation, acupuncture, relaxation, therapeutic hypnosis, tai chi, qi gong, Pilates, Feldenkrais method); and other approaches (e.g., traditional Chinese medicines, Ayurvedic medicines).

The Cochrane Collaboration, under the leadership of American academics Susan Wieland and colleagues, chose to rely on an operational rather than a theoretical definition, while retaining the generic name CAM. Here again, due to the complexity, the authors decided to empirically exclude most psychotherapies ("with the exception of unconventional psychotherapies"), most physical activity programs for therapeutic purposes ("at the exception of mind-body practices"), and food and vitamin supplements from their field of interest (Wieland et al. 2011). These restrictions lead to systematic reviews and truncated meta-analyses in many care or prevention activities.

Isabelle Boutron and Philippe Ravaud argued that non-pharmacological treatments "concern technical interventions such as surgical procedures; technical as stents and arthroplasty; non implantable devices such as orthoses, ultrasound treatments, and laser treatments; and participative interventions such as rehabilitation, education, behavioral intervention, and psychotherapy" (Boutron et al. 2012, xi). The fourth edition of the book *Integrative Medicine* authored by David Rakel and colleagues differentiated heterogeneous categories such as physical modality, biomechanical therapies, supplements, nutrition, mind-body therapy, botanicals, aromatherapy, bioenergetic therapies, traditional Chinese medicine, Chinese herbs, general measures (or lifestyle change), and homeopathy (Rakel 2018).

Several websites list practices according to a medical specialty (gastroenterology, for example) or to a direct economic interest; complementary medicines represent a substantial, growing industry with manufacturing jobs and export potential. Organizations sort practices in various ways; they include or exclude techniques according to their beliefs and culture.

As long as one is interested in the subject from afar, all of these practices may seem similar. From a proximal viewpoint, however, it feels like an elusive world, and regretting the heterogeneity that comprises the complexity of the field serves for nothing, except impostors. The time has come to answer the question: is this practice an effective and safe solution for this health problem? The time has come to characterize, hierarchize, and qualify all of these practices. A transparent and rigorous methodology is required to determine their properties. The next section is dedicated to this goal.

False Approaches, False Defenses, Real Limitations

Between ingredient and method, between alternative and complementary medicine, between intuitive and evidence-based methods, between public health messages and individual health solutions, and between diagnostic and care methods, the time has come to know what we are talking about.

From a science viewpoint, the time also has arrived to come to an easy conclusion such as "the evidence is not conclusive because of high levels of heterogeneity, publication bias, and the risk of bias in the majority of studies."

4   To Get out of the Nebula with the STRIACOD

4.1   Pseudoscience or Science

Let us get rid of ambiguity. Some alleged "health" methods are based on pseudoscience, and they claim exemption from assessment or scalable processes. Pseudoscience methods are hazardous, intuitive, and singular (not reproducible). Some refer to unexplainable processes (e.g., magic light, invisible energy, traditional ritual). Some attribute effects to the professional (e.g., guru, master). Some attribute effects to the context (e.g., room, town). These intuition-based, preference-based, and/or traditional practices cannot be evaluated because they belong to ritual and supranatural effects. A bonesetter, an herbalist, a scraping and cupping therapist, a village healer, a witch doctor, a faith healer, and a naturopath use esoteric and mystic explanations. Philosophical and lifestyle approaches have to be efficient; otherwise, it is as nonsensical as trying to prove the existence of God. Esoteric, mystic, and supranatural approaches are not medicines. An individual practice of religion is respectable. Attempting to verify evidence of that practice's health benefits and risks in rigorous clinical trials would be stupid. These communities do not want evidence because they have belief. Whether it be truth, fake, or explanation, religion and philosophy are components of cultural education and social approaches that are unverifiable by comparative studies. They reflect community and political decisions.

On the opposite side, some health methods are fundamentally based on science, such as drugs. Their protocols for human health are evidence-based, science-based, scalable (constant improvement), universal, programmable, describable (concrete), timed (before and after), known (benefits, risk, %), and recommended by health professionals (e.g., prescribed by generalist or specialist medical practitioners). They are administered on a personalized basis by a licensed professional operating in coherence with a medical diagnosis and they are monitored and adjusted accordingly. They follow the WHO (2019b) definition of health interventions, without the term "assess" (see 4.4): *"A health intervention is an act performed for, with or on behalf of a person or population whose purpose is to, improve, maintain, promote or modify health, functioning or health conditions."* (WHO 2019b). Protocols can be hierarchized with the accumulation of data and differentiated from unproven and/or dangerous methods (Table 1.5). There is no reason to think that some proto-

Table 1.5 Unproven and/or dangerous methods

Unproven Methods	Dangerous Methods
Intuitive practices	Sectarian practices (sentenced by justice)
Traditional protocols approaches	Dangerous practices (alerted by a health authority)
Alternative medicines	

cols would not be better and safer than others for certain patients. Science verifies hypotheses with experimentation. Positive and negative results need to be published. New evidence will modify a statement with a better strategy, and so on.

Science is one of the cornerstones of NPIs. There are science-based protocols with convergent and consistent effects in humans. There are also protocols that are revocable with the appearance of new evidence.

4.2   Test or Treatment

Let us get rid of ambiguity. Scientifically, assessment tests require validation methodologies to distinguish between positive and negative subjects (e.g., AIDS, COVID-19) or to determine the value of a variable (e.g., blood pressure, symptom intensity). Data are collected, verified, compared to norms, and linked to health classifications such as, the International Classification of Disease (ICD); the *Diagnostic and Statistical Manual of Mental Disorders* (DSM); or the International Classification of Functioning, Disability and Health (ICF). Their specificity, sensitivity, fidelity, and accuracy are the main statistical characteristics. Thus, these tools are dedicated to determining a medical diagnosis: for paramedical evaluation and for assessing body composition, health behaviors, cognitive function, education gaps, social needs, disability, impairments, and treatments, such as non-pharmacological practices. They contribute to determining a strategy, for example, tools such as a neuropsychological test, a body sensor, or a digital tool for counting steps are not methods that change health status when used alone.

On the opposite side, an intervention with a clear goal of improving health (cure, care, prevent, optimize health parameters), a protocol (ingredients, sessions), and a predefined period meets the condition to be a non-pharmacological practice. The impact of the intervention needs to be verified, and this verification requires a before-after intervention study (Chap. 8). The results of these experiments need to consistently confirm hypotheses of positive health effects and the absence of secondary effects. Treatment assessments take a long time to obtain and their results are counterintuitive; often they are disappointing when compared to the observations provided by exploratory, qualitative, observational, or in vitro methods. This condition is the second cornerstone of NPIs.

4.3   Ingredient or Recipe

Let us get rid of ambiguity. Natural products such as roots, fruits, mushrooms, vegetables, seeds, plants, algae, and barks may have health-promoting properties (see example below). Their reputation, however, is not enough; there is a big step needed in order to go from reputation to making health recipes. The association of a food product to a health condition is not a sufficient condition to guarantee health benefits. Beyond controlling the quality of the product, its potential health benefits depend heavily on the preparation, the quantity consumed, the time at which it is taken according to chronobiology, and the duration for which it is taken.

Ginger
This root facilitates weight loss, lowers bad cholesterol, is an antispasmodic, relieves heartburn, decreases nausea, is an antiviral, is an antioxidant, is an antiseptic, is an anti-inflammatory, is an aphrodisiac, and boosts energy.
Guarana
This fruit fights against intestinal disorders, decreases muscle pain, boosts the immune system, fights against migraine, stimulates the body, is euphoric, is diuretic, is a pain reliever, boosts energy, increases memory, promotes digestion, and detoxifies blood.
Shiitake
This mushroom is an antioxidant, boosts the immune system, boosts metabolism against diabetes, lowers blood cholesterol levels, hardens arteries, soothes eczema, treats colds, treats prostate or breast cancer, treats herpes, and decreases high blood pressure.
Spinach
This vegetable is anticancer, improves the beauty of hair, strengthens bones, improves visual acuity, cuts hunger, fights against asthma, decreases anemia, lowers blood pressure, and boosts the digestive system.
Pistachio
This seed helps fight stress, lowers cholesterol, fights cardiovascular disease, improves eyesight, protects skin and hair, and is an antioxidant.
Moringa
This plant helps fight diabetes, strengthens the immune system, balances cholesterol, is an anti-inflammatory, fights constipation, is an antioxidant, decreases appetite, detoxifies, prevents cancer, is antifungal, is an antidepressant, stimulates hair growth, and improves libido.
Spirulina
This alga stimulates the immune system, acts against thyroid disorders, is an antioxidant, helps lower cholesterol, lowers blood pressure, helps fight against anemia, reduces the risk of diabetes, improves kidney function, reduces fatigue, helps weight loss, and promotes muscle growth.
Cinnamon
This bark is an antioxidant, reduces blood sugar levels, promotes weight loss, helps fight fatigue, stimulates blood circulation, strengthens metabolism, is an aphrodisiac, and boosts energy.

On the opposite side, a recipe needs to accurately combine all ingredients in coherence with an approach and a professional discipline (Table 1.6). In the margins, it is worth noting that there can be differences in these combinations depending on individual culture, context, and tastes. However, the expected effect can only be achieved by following a strict and sequential protocol; preparation is a key element.

Table 1.6 Position of an NPI from the global scale to the concrete scale

Scale	Detail	Example
Approach	Point of view of human complexity	Psychology
Discipline	Professional strategy	Psychotherapy
NPI	Method, protocol, manualized intervention	MBSR
Technique	Ingredient	Mental body scan
Material	Practical element	Practice log

A pizza is not just made up of tomatoes. While the ingredients are common, there may be variations in their combinations. To be effective, the steps of production are sequenced over time, always in the same order. This reasoning for natural ingredients is the same for any single skill, manual gesture, and sport material.

Without protocol, no non-pharmacological practice can be available and relevant. Non-pharmacological practices need to have universal and reproducible health impacts. To use treatments tested in trials, clinicians need sufficient details of the "how to." Many current trials and reviews often omit crucial details of treatments. Providing some additional treatment details could improve the uptake of trial results in clinical practice. Some international integrity guidelines for professional and training sessions are developed, for example, in the field of mindfulness-based stress reduction (MBSR) programs (Kenny et al. 2020). The guidelines ensure that professionals are consistent in offering their participants with appropriate and evidence-based skills and frameworks. The guidelines advertise and offer information and ensure that non-pharmacological recipes are grounded in ethical considerations (e.g., integrity, transparency, openness, relational responsibility, clarity around limitations).

The "Who, What, when, and where" of the Treatment

Glasziou et al. (2008) prospectively assessed 80 published reports of treatment from October 2005 to October 2006 (55 single randomized trials, 25 systematic reviews). Articles were published in the *New England Journal of Medicine* (10), *Cochrane Database of Systematic Reviews* (9), *Lancet* (7), *JAMA* (7), *Archives of Internal Medicine* (6), *BMJ* (5), *Annals of Internal Medicine* (5), and several other journals (31). The treatments, drugs, or NPIs needed to be highly relevant to clinical practice. Less than half (36/80) were NPIs: education and training (15), devices or surgery (10), psychological treatments (4), service delivery (3), and a mix of other interventions (4). The description was satisfactory for 29% of studies on NPIs while it was 67% in drug studies. The authors concluded that "information was better in reports of individual trials than in systematic reviews, and for drug treatments than for non-drug treatments (p.1472)". Another study confirmed that 39% of 137 interventions assessed were described in sufficient detail in the published primary report to enable replication of the intervention in practice (Hoffman et al. 2013). International guidelines such as CONSORT recommend ensuring that the primary paper makes explicit mention of all related documents (such as protocols, online supplementary material, and websites). If researchers build complete description of the intervention, clinicians and patients can reliably implement useful interventions.

4.4 Alternative or Integrated

Let us get rid of ambiguity. Semantics are important. An alternative medicine is supposed to be secret, and their use is frequently under-declared to doctors (Lognos et al. 2019). An alternative medicine is supposed to replace a conventional solution; yet, the main alternative medicines claim to have a holistic approach to humans. How can we integrate several holistic approaches without misunderstanding? Cheirology is an art of hand analysis, a combination of ancient Chinese Buddhist hand analysis and the best of traditional Western palmistry. How can we combine cheirology, massage therapy, physiotherapy, and exercise therapy in a relevant dialogue? Fusion creates confusion. A medicine that is conceived of as integrative, the result of the integration of several treatments and preventives strategies, will be relevant. On the other hand, an approach that merely juxtaposes several integrative medicines will be hazardous at best, dangerous at worse.

What are alternative treatments exactly? Are they only the non-studied, unproved, noneffective, unsafe, or unregulated? *"Alternative medicine describes any practice that aims to achieve the healing effects of medicine, but which lacks biological plausibility and is untested, untestable or proven ineffective."* (Wikipedia 2020). Some examples are traditional medicine or energy medicine, faith healing, herbalism, shamanism, and naturopathy.

On the opposite side, the vocation of non-pharmacological practices is to be transparently integrated in the health course curricula of each individual. It is not a secret. It is not a practice related to some underground world. The goals of its practitioners are to take part in a therapeutic process. Non-pharmacological practices must be integrated, tracked, controlled, and covered by insurance. This integration should follow a strategy of complementarity that makes it concerted, monitored,

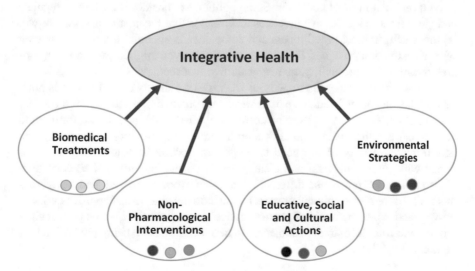

Fig. 1.1 Integration process from different solution components

and adjusted if necessary. Some practices claim to be holistic. Logically speaking, it is thus impossible to associate two holistic approaches; doing so would be confusing and sometimes counterproductive. Combining antidepressant drug and exercise therapy in a patient with moderate depression has no benefits if it is done alone. If the integration of evidence-based interventions is a part of a pragmatic process that is carried out in a network by professionals working together in a multidisciplinary team, then the chances of success will be maximal. It is by taking the best of each practice and considering their possible interactions that we will be able to fight against health problems that are not resolved by biomedical therapies, such as the control of noncommunicable diseases, cancer care, well-aging, and the end of life.

As such, the integration process differs from its solution components. Integration takes solutions from biomedical therapies, NPIs, sociocultural approaches, and environmental approaches in order to propose the best solutions for individuals, depending on their motives, context, and health conditions (Fig. 1.1). This all-inclusive offer is currently far from being practiced in the majority of countries.

4.5   Used or Approved

Let us get rid of ambiguity. For the moment, few institutions have approved or disapproved non-pharmacological practices. Debates are heated between experts, medical societies, authorities, politicians, and between health insurance companies. No official or consensual label exists as of yet. Practitioners do not have to follow any specific procedures. No serious regulatory texts exist for an approved or disapproved practice, except for a few condemnations. Consumers are calling, with some gullibility, for a complete freedom of use. Acceptance and authorization processes remain optional in most countries. Some politicians consider it useless to regulate and control these inoffensive practices and assume that the cost of assessment will be too much, trusting practitioners and companies is enough, and that the market will select the best practices. Faced with complexity, you might as well do nothing – no evaluation, no traceability, no responsibility. Not seen, not caught.

On the opposite side, the tragedy of COVID-19 reminds us that health is not a market like any other. Medical, paramedical, and prevention protocols must be clear and rigorous. Their use needs to be correctly applied to the population for them to have a real health impact. The effects on human health, autonomy, longevity, and health-related quality of life need to be patent. Accumulated and congruent evidence will give a universal and clear position on the benefits and risks of non-pharmacological practices, differentiating them from unproved and dangerous methods. Moreover, a national or supranational organization must approve or disapprove practices. In the world of pharmaceutical drugs, a product can be permanently withdrawn from the market or subjected to new studies before being put back on the market (Fig. 1.2).

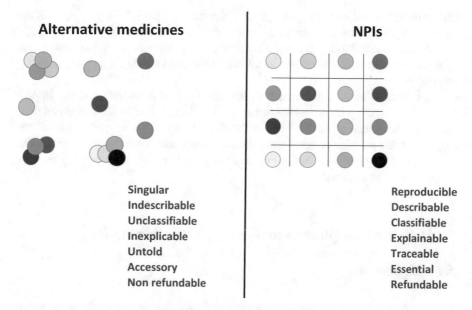

Alternative medicines

NPIs

Singular	Reproducible
Indescribable	Describable
Unclassifiable	Classifiable
Inexplicable	Explainable
Untold	Traceable
Accessory	Essential
Non refundable	Refundable

Fig. 1.2 Difference between NPIs and alternative medicine

4.6 Decided or co-Decided

Let us get rid of ambiguity. Practitioners have essentially learned by oral and imitative processes. The application of the conviction and the persuasiveness of professionals over patients who are weakened by a disease are frequent. Facing a lack of information and limited time for consent, patients can accept a practice without understanding the limits of its benefits, its potential side effects, the risks of its interaction with other treatments, constraints on its implementation, its time burden, any short- and medium-term costs it may incur, or even its indirect cost.

On the opposite side, non-pharmacological practices need to be codified, both for their professionals and their patients. Users need to be fully and freely informed and in a condition to accept their use – consent is essential. Being participative and proactive to a treatment is a non-neglected part of its effect. Making a co-decision with a team of health professionals should be preponderant, as should be the right to abandon any aspect of that treatment.

4.7 Indirect or Direct Effect

Let us get rid of ambiguity. Health is not the main goal of work, sport, or transportation; incidents, injuries, or accidents are there to remind us of this fact. Sociocultural activities such as cinemas, concerts, or nightclubs do not prioritize a positive impact on the overall health of their participants in a short-term period. Furthermore, in the short-term positive environmental actions such as a river cleaning, the creation of a roof garden, a fun development of a staircase, or a reduction of the number of cars in a city are only directly related to a health change in a specific person or group of people.

On the opposite side, non-pharmacological practices must have a direct impact on subjects' health within a 3-month period. This impact can be demonstrated with a before-after intervention study that determines the reality and systematicity of the impacts of that practice on measurable health parameters. The size of this effect can be compared to other interventions (pharmacological or not). The causality relation needs to be demonstrated.

5   Defining Non-pharmacological Interventions (NPIs)

5.1   Definition

To get out of this nebula, non-pharmacological practices have to follow the seven previous conditions named STRIACOD: science, treatment, recipe, integration, approbation, co-decision and direct effect. Scientists have proposed calling the health solutions meeting these criteria "non-pharmacological interventions" with the acronym NPIs. The can be preventive and/or therapeutic solutions, but they must be able to be both described precisely and rigorously evaluated. Let us operationally define this concept.

> **Definition**
> According to the Academic Collaborative Platform, Plateforme CEPS in 2017, "*NPIs are science-based and non-invasive interventions on human health. They aim to prevent, care, or cure health problems. They may consist of products, methods, programs or services whose contents are known by users. They are linked to biological and/or psychological processes identified in clinical studies. They have a measurable impact on health, quality of life, behavioral and socioeconomic markers. Their implementation requires relational, communicational and ethical skills*". (Plateforme CEPS 2017).

Non-pharmacological interventions are supervised solutions that have been proven by studies demonstrating their benefits on health, autonomy, longevity, and/ or health-related quality of life, and on the identification of health risks (side effects, dangerous interactions). To be designated as an NPI, a description of the intervention and its main attributes is required (Table 1.7). It is a sine qua non condition.

Table 1.7 Conditions to be an NPI

	Obligatory	Optional
Designation	Last name	Acronym, synonym(s), author(s), institution(s), label(s)
Health goal	Main health problem to prevent, treat, or cure	Secondary benefits
Target population	Minimum–maximum age	Gender, socio-educational level, territory
Content	Components (ingredient, technique, gesture), procedure (duration, session, dose), materials	Precaution, professional manual, user manual
Context	Place of practice, time of the health pathway	Medical prescription (or not), reimbursement
Mechanisms	Main mechanisms	Secondary mechanisms
Provider	Operator profession	Initial training diploma, continuing education diploma, qualification, certification
Scientific publications	$\geq$ 1 publication of a positive intervention study	Other publications (other methods, systematic reviews, collective expertise, etc.)

Complementary description can be provided as a means of reinforcing its credibility and dissemination. NPIs are most often delivered or given to patients by a trained human.

These practices are extensively used in prevention. In treatment, they should only very rarely be used as therapeutic alternatives. NPIs complement the authorized biomedical treatments approved by the U.S. Food and Drug Administration (FDA) in the United States or by the European Medicines Agency (EMA) in Europe. They are frequently used in the practice of general medicine, internal medicine, gerontology, integrative medicine, oncology, rehabilitation, psychiatry, preventive medicine, and palliative medicine. They follow the diagnostic precepts of modern Western medicine. NPIs engage patients so that they become proactive in the adoption of new health behaviors. They involve patients so that they can foster self-management and empowerment.

5.1.1 The Choice of NPIs

The academic platform CEPS team has long sought a positive and inclusive synonym for naming NPIs (Ninot 2013). Discussions between professionals and users and meetings of experts were organized for this purpose between 2014 and 2015. A pilot survey was proposed in 2015. A national survey was conducted during the iCEPS International Conference in March 2015 in Montpellier, France. Finally, the term "non-pharmacological intervention" obtained the maximum number of votes. Why?

> **The *Plateforme CEPS***
> The "*Plateforme CEPS*" was created in 2011 at the University of Montpellier in France. This collaborative academic platform studies relevant methods for assessing the benefits and risks of NPIs for health, autonomy, longevity, and health-related quality of life. Its work is conducted with the key-players of the health ecosystem, including practitioners, preventionists, social workers, caregivers, users, researchers, decision-makers, young or old, sick or not sick. The platform currently includes 140 collaborators from 11 different countries, and it offers a list of methodological invariants for any study evaluating an NPI (www.plateforme-ceps.fr). In 2011, the platform created the iCEPS Conference, an international scientific congress on NPIs, www.icepsconference.fr. The platform has also developed, with the support of academic partners (Europe, State, Occitanie Region, Montpellier Metropole, French National Cancer Institute, SIRIC Montpellier Cancer, CARSAT Languedoc-Roussillon, Occitanie Regional Health Agency), and Open Science digital systems intended to help those evaluating NPIs:
>
> - A meta-search engine for publications of clinical and intervention studies on NPIs called *Motrial:* www.motrial.fr:
> - A library providing free access to publications of clinical and intervention studies on NPIs targeted to cancer prevention and therapy called *Inmcancer*: www.inmcancer.fr
> - A library providing open access to clinical and interventional studies evaluating NPIs for healthy and successful aging called *Bienvieillirinm*: www.bienvieillirinm.fr
> - A system for sharing academic resources on NPIs such as master's degree reports, PhD reports, unpublished research reports, and slideshows called *Nishare*: www.nishare.fr
> - An international directory of institutions and researchers specialized in the evaluation of NPIs named *Niri*: www.niri.fr

5.1.2   The Term "Non"

This means that the operative process of NPIs is never invasive like medical surgery, transplantation, medical device implantation, gene therapy, or intravenous injections.

5.1.3   The Term "Non-pharmacological"

This means that the action mechanisms of NPIs are non-pharmacological, instead calling on other biological, behavioral, and/or psychosocial processes. Using the word "pharmacological" instead of drug would have placed too much emphasis on the product and not the process. The use of the term "pharmacological" pays homage to the rigorous process of clinical validation and surveillance adapted by academics and industrials. It departs from the word "alternative," which suggests esoteric practices or fake medicines refusing science and quality assessments. It also departs from the word "complementary," which gives the impression of a "secondary," "accessory," or "optional" intervention. It departs from the word "integrative," which is too vague, amalgamating decision and solution process without the official recognition needed to obtain legal and financial support. NPIs are intended to be conventional medicines in their own right; therefore, they are validated, and according to relevance, prescribed and monitored by a doctor.

5.1.4   The Term "Intervention"

It covers the idea that a human will have a specific action following a precise and timed method for prevention, care, or cure. A codified protocol needs to be available both for professionals and users. According to ClinicalTrials, an intervention (or treatment) is "*a process or action that is the focus of a clinical study. Interventions include drugs, medical devices, procedures, vaccines, and other products that are either investigational or already available. Interventions can also include noninvasive approaches, such as education or modifying diet and exercise*" (ClinicalTrials 2020).

NPIs Are
- Applied to solve a targeted health problem (a symptom, a disease, a risky behavior)
- Validated by rigorous studies published in independent peer-reviewed scientific journals
- Time-limited protocols
- Individual and group methods
- Personalized approaches considering the context and lifestyle of each individual
- Supervised by a specifically trained professional
- Codified methods bringing together several mandatory specifications
- Freely consented to by the user after being fully informed by the professional
- Combined protocols with conventional medicines if necessary

NPIs Are Not.
- Lifestyles inspired by a philosopher or a community leader.
- Public health campaigns promoting positive health behaviors (e.g., active mobility, eating fruits and vegetables) or discouraging risky behaviors (e.g., smoking, drinking, fast-fooding, sedentary behavior)
- Environmental or architectural solutions that seek to influence the health of a population over a long term (e.g., bike paths in towns, river clean-ups)
- Organizational health solutions (e.g., digital cards sharing personal health data between professionals)
- Diagnostic tools (e.g., body sensors, pedometers)
- A single ingredient (e.g., plant, vitamin, mineral), technique (e.g., posture, massage gesture, breathing skill), or material (e.g., massage accessory, digital chair)

5.1.5   Use of the NPI Term

Today, this term is widely used in the literature (Boutron et al. 2012; Ninot 2013, 2019). Academic institutions, medical societies, and states use it as well. If you try, you will find non-pharmacological interventions using an Internet browser. Most of all, this concept is recognized and understood by users. Chances are, in the long run, the acronym NPI will prevail over the full appellation; however, staying on a macroscopic level will not allow you to understand each individual NPI. It is not the name of the toolbox that matters, but each tool it contains and the instructions for their use. The challenge is to highlight the specificities of each method for each health problem. Who, what, when, and where? Full descriptions of treatment must include procedures used, the timing of treatment, including duration and intervals of dosing or sessions, materials needed, and accessibility of any instructions including overcoming language barriers (Glasziou et al. 2008). Their number seems infinite and will only increase as research progresses, technology advances, and as new diseases appear. Drugs have made their revolution; it is now the turn of NPIs.

Should there be concern over the use of a negative appellation, a positive term linked to the intrinsic properties of NPIs and their relation to human health interventions can be used, such as "human and natural health methods" (HUNAMs).

Find Relevant Study Articles

To search for studies or systematic reviews evaluating the benefits and risks of NPI on health, you must look on a scientific search system such as PubMed, Motrial, Embase, Kalya Research, Science Direct, Google Scholar, or Central. Use at least one of these words, and if possible all eight, in order to obtain the most exhaustive query possible: "non-pharmacological OR nonpharmacological OR non-pharmaceutical OR nonpharmaceutical OR nonpharmacologic OR nonpharmacologic OR non-drug OR nondrug."

The term "intervention" can be deleted in the query. Authors of studies may indifferently use approach, management, method, practice, process, protocol, therapy, recipe, remedy, strategy, solution, technique, therapy, or treatment, in either the singular or plural. Beware of the terms "alternative," "complementary," "integrative," "natural," or "traditional," which will give you imprecise and biased results.

The reference search must include all the English synonyms and the abbreviations of the NPI. Databases such as PubMed or Science Direct are not sufficiently exhaustive on NPIs. Google Scholar is too large. Embase is not free of charge.

5.2   Goals of NPIs

Each NPI aims to solve a primary, targeted health problem:

- Improve a health behavior such as sedentary behaviors or drug compliance.
- Prevent health troubles such as lung cancer or obesity.
- Treat symptoms such as pain or fatigue.
- Delay disease aggravation such as in cardiovascular disease or a cancer.
- Cure diseases such as type 2 diabetes or early-stage depression.

NPIs will have secondary, positive impacts on a patient's overall health condition, autonomy, longevity, and an increase in their health-related quality of life.

For an individual, beyond the benefits on health indicators, the use of NPIs improves their capacity to maintain social participation (e.g., work), reduces family dependence, and saves money (e.g., direct medical costs). In all cases, the use of an

NPI makes the patient an actor in their own care by giving them the conviction that they have all the chances on their side; at least to get better, and at most to heal. In other words, they are empowered.

Optimize NPI Use with Evaluation
Researchers make preventive and therapeutic interventions better. They must, henceforth, help healthcare professionals, prevention actors, and patients to answer the following questions for each NPI:

- How does it work?
- What are the psychological processes and biological mechanisms at play?
- What is its therapeutic or preventive indication?
- What health problem can be solved?
- What is the content of the practice?
- What dosage?
- What duration, intensity, frequency?
- What actions should be taken and with what precautions?
- What are the risks of interference and interaction with biomedical treatments?
- What are its undesirable effects?
- What are its contraindications?
- Who are its professionals with which initial and continuing training?
- Who is to prescribe, monitor, supervise, and adjust its use?

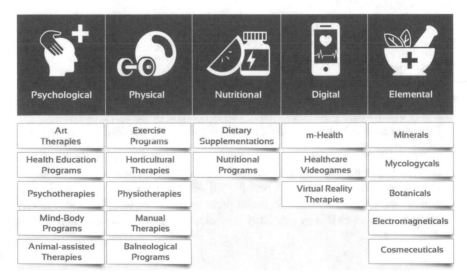

Fig. 1.3 NPIs classification. (Published with kind permission of © Plateforme CEPS. All rights reserved)

For society, NPIs will decrease the burden on the health system (e.g., non-programmed treatments, visits, consultations, examinations, hospitalizations, transportations) and production losses (e.g., unemployment contributions, social assistance). The NPI field creates jobs in care, in prevention, in personal assistance services, in innovation, and in research. It is one of the growing markets of the future (see Chap. 7).

5.3   Classification of NPIs

A vague title, a professional discipline, or a technique that says nothing about the exact content of an NPI, its implementation procedure, or its health objective is cause for the non-classification of an NPI. For example, the term "naturopathy" is too vague to give a clear idea of what methods are employed by the professional using it to treat patients. The term "clinical psychologist" says nothing about the psychotherapies mastered by the professional. A gesture, like acupressure, says nothing about the manual therapy used. An ingredient, like honey, says nothing about a proposed food supplement.

It is necessary to describe the NPI, its method, its theories, its mechanisms of action, its content, its dose, and the context of its use. This transparent description applies as much to clinicians and users as it does to researchers and policy makers.

A description including the science and the quality of each NPI approach and using a common base will allow them to be better identified, to be better categorized, to be better compared, and to better define both their interests and limits for health. This will improve practices, allowing some of them to be used with more frequency and to be better reimbursed.

A classification emerges through a co-constructive process involving professionals, patients, and researchers. Like any classification, it will never be definitive (Fig. 1.3).

> **NPI Categories and Subcategories**
> The Plateforme CEPS differentiates five categories – psychological, physical, nutritional, digital, and elemental health interventions – and twenty subcategories. This classification offers a better representation of NPIs belonging to each subcategory.

NPIs are, therefore, preventive actions or therapies that have been "manualized" (described in a specification document), and whose benefits and risks have been identified by studies published in peer-review scientific journals. In this sense, they are medicines in their own right, sometimes to be used in complement to biomedical treatments, sometimes to prevent, and sometimes to be used as recourses following biomedical failure (e.g., palliative care).

Unlike drugs and medical devices, NPIs do not currently have any legal status, whereas a drug follows strict regulations from its scientific validation to its market access, and from its production to its use (Chap. 8).

6   Defining NPI Categories and Subcategories

6.1   Psychological Health Interventions

"Psychological health interventions" have proximal related terms such as psycho-educational health interventions, psychosocial health interventions, social health interventions, public health interventions, or population-based health interventions. "Psychological health interventions" refer to the 2017 WHO definition:

> Psychological interventions that are potentially scalable include modified, evidence based psychological treatments, such as:

- Brief, basic, non-specialist-delivered versions of existing evidence-based psychological treatments (e.g., basic versions of cognitive-behavioural therapy, interpersonal therapy);
- Self-help materials drawing from evidence-based psychological treatment principles, in the form of self-help books, self-help audiovisual materials, and online or app-based self-help interventions;
- *Guided self-help in the form of individual or group programs, providing people with guidance in using the above-mentioned self-help materials.*

This category includes five subcategories:

The first subcategory relates to "art therapies", with proximal related terms such as artherapies, art-based health methods, and art-related health programs. In 2013, The American Art Therapy Association defined them as follows:

> Art Therapy is an integrative mental health and human services profession that enriches the lives of individuals, families, and communities through active art-making, creative process, applied psychological theory, and human experience within a psychotherapeutic relationship.

Examples include music therapy, plastic modelling therapy, dance therapy, and painting therapy. Professionals can be called art therapists, music therapists, and dance therapists.

The second subcategory is "health education programs", with proximal terms such as educational programs, health education methods, psychoeducation programs, community-based programs, health education strategies, disease management programs, and specific public health programs. In 2020, the WHO defined them as:

> Health education is any combination of learning experiences designed to help individuals and communities improve their health, by increasing their knowledge or influencing their attitudes (WHO website 2020a, b).

Examples include primary prevention interventions, secondary prevention interventions, tertiary prevention interventions, and focused public health programs. Professionals can be nurses, physicians, educators, social workers, coaches, case managers, doctors, pharmacists, HIV counsellors, family planning counsellors, addiction counsellors, bereavement counsellors, clinical social workers, district social welfare officers, sexual assault counsellors, women's welfare organizers, midwives, mental health support workers, and health vocational education teachers.

The third subcategory is "psychotherapies" with proximal related terms such as psychological therapies, talk therapies, or psychotherapeutics. In 2019, The American Psychiatric Association defines them as follows:

Psychotherapy, or talk therapy, is a way to help people with a broad variety of mental illnesses and emotional difficulties. Psychotherapy can help eliminate or control troubling symptoms so a person can function better and can increase well-being and healing. Problems helped by psychotherapy include difficulties in coping with daily life; the impact of trauma, medical illness or loss, like the death of a loved one; and specific mental disorders, like depression or anxiety. There are several different types of psychotherapy and some types may work better with certain problems or issues. Psychotherapy may be used in combination with medication or other therapies (American Psychiatric Association website 2019).

Examples include cognitive behavioral therapy, interpersonal therapy, dialectical behavior therapy, psychodynamic therapy, psychoanalysis, and supportive therapy. The professionals are psychotherapists, clinical psychologists, psychiatrists, and family therapists.

The fourth subcategory is "mind-body programs," with proximal related terms such as body-mind programs, psychosomatic therapies, and mind-body methods. A definition proposed by the National Cancer Institute in 2020 states:

A health practice that combines mental focus, controlled breathing, and body movements to help relax the body and mind. It may be used to help control pain, stress, anxiety, and depression, and for overall health (NIH website 2020).

An example is yoga. The professionals are psychotherapists, psychiatrists, mind-body practitioners, licensed professional counselors, physiotherapists, psychomotor practitioners, nurses, and midwives.

The fifth subcategory is "animal-assisted therapies," with proximal terms such as zootherapies and traditional healing with animals. A definition proposed by the American Veterinary Medical Association in 2020 states:

Animal-assisted therapy (AAT) is a goal directed intervention in which an animal meeting specific criteria is an integral part of the treatment process. Animal-assisted therapy is delivered and/or directed by health or human service providers working within the scope of their profession. Animal-assisted therapy is designed to promote improvement in human physical, social, emotional, or cognitive function. Animal-assisted therapy is provided in a variety of settings, and may be group or individual in nature. The process is documented and evaluated.

Professionals are zootherapists and nurses.

6.2   *Physical Health Interventions*

The category of "physical health interventions" can be defined as *"an individual or group, non-invasive, manualized, supervised and comprehensive program using passive or active mobilization of the body improving significantly health markers with a patently impact on health optimization, prevention or care."*

The first subcategory includes "exercise programs" with proximal related terms such as physical activity programs, adapted physical activity programs, sport therapies, health training programs, and fitness programs, and is defined as follows by the WHO in 2020 (WHO 2020c):

> *Exercise is a subcategory of physical activity that is planned, structured, repetitive, and purposeful in the sense that the improvement or maintenance of one or more components of physical fitness is the objective.*

An example of a classified practice is the Otago program. Professionals are adapted physical activity educators, physiotherapists, fitness instructors, and sport coaches.

The second subcategory is "horticultural therapies" with proximal related terms such as garden therapies. In 1997, The American Horticultural Therapy Association (AHTA) defined horticultural therapy as:

> *Horticultural therapy is the participation in horticultural activities facilitated by a registered horticultural therapist to achieve specific goals within an established treatment, rehabilitation, or vocational plan. Horticultural therapy is an active process which occurs in the context of an established treatment plan where the process itself is considered the therapeutic activity rather than the end product.*

Professionals are horticultural therapists and nurses.

The third subcategory is "physiotherapies" with proximal related terms such as physiotherapy protocols, kinesitherapies, or movement therapies. In 2020, The Chartered Society of Physiotherapy defined physiotherapies as:

> *Physiotherapy helps restore movement and function when someone is affected by injury, illness or disability.*

An example is the Mezieres method. Professionals are physiotherapists, geriatric physical therapists, orthopedic physical therapists, pediatric physical therapists, chiropractors, osteopaths, kinesithcrapists, and kinesiologists.

The fourth subcategory is "manual therapies" with proximal related terms such as touch therapies, manual therapy techniques, orthopedic manipulative physical therapies, massage therapies, massage programs, manual treatments, or functional manual therapies. In 2014, Clar et al. defined manual therapy as follows:

> *A non-surgical type of conservative management that includes different skilled hands/fingers-on techniques directed to the patient's body (spine and extremities) for the purpose of assessing, diagnosing, and treating a variety of symptoms and conditions. Manual therapy constitutes a wide variety of different techniques which may be categorised into four major groups: a) manipulation (thrust manipulation), b) mobilisation (non-thrust manipulation),*

c) static stretching, and d) muscle energy techniques. The definition and purpose of manual therapy varies across health care professionals.

Examples of manual therapies can be healing touch, massage, manual technique, hands-on techniques, and therapeutic touch. Professionals are physical therapists, physiotherapists, occupational therapists, chiropractors, massage therapists, athletic trainers, and osteopaths.

The fifth subcategory is "balneological programs" with proximal related terms such as balneological methods, thermalism programs, thermalism cures, hydrotherapies, medical hydrology and climatology techniques, spa treatments, or spa therapies. In 2020, the Balneology Association of North America defined balneology as follows:

The study of the art and science of baths and bathing in natural mineral waters for health and wellness purposes. Balneology includes scientific research into the methods and applications of bathing, drinking, steaming and inhaling natural thermal and mineral waters for wellness, health and medical benefits, including their associated natural gases and peloids (organic muds). As a wide and interdisciplinary field, it encompasses not only site-specific analyses and clarifications of natural mineral water sources, but also their local and regional geography and geology, mineralogy and chemistry, bio-molecular nature and structural forms, and climates and seasons.

Professionals are physiotherapists, manual therapists, hydrologists, and nurses.

6.3 Nutritional Health Interventions

The nutritional health interventions category was defined by the WHO in 2017 as follows:

Nutrition is a critical part of health and development. Better nutrition is related to improved infant, child and maternal health, stronger immune systems, safer pregnancy and childbirth, lower risk of non-communicable diseases (such as diabetes and cardiovascular disease), and longevity.

The first subcategory is dietary supplementations with proximal related terms such as supplements, dietary supplements, enriched foods, food supplements, functional foods, medicinal foods, novel foods, and nutraceuticals. In 2018, the World Cancer Research Fund defined it as:

A dietary supplement is a product intended for ingestion that contains a 'dietary ingredient' intended to achieve levels of consumption of micronutrients or other food components beyond what is usually achievable through diet alone.

Examples include nutritional supplements, enriched foods, fortified foods, or novel foods. Professionals are dieticians, clinical dieticians, nutritionists, nutritional therapists, pharmacists, food service dieticians, nutritionists, sports nutritionists, and sports dietitians.

The second subcategory is nutritional programs, with proximal related terms such as diets or nutritional therapies. In 2020, the Institute of Health Sciences defined it as:

> *Nutritional therapy practically applies the latest theories and research in nutrition and health sciences to individuals seeking to manage chronic disease or promote optimum health. The focus is improving physiological function of a number of bodily systems, including the digestive, immune, endocrine and cardiovascular systems. Nutritional Therapy can help alleviate and manage a wide range of conditions and can help improve the health outcomes of many individuals.*

Examples include diets. Professionals are dieticians, clinical dieticians, nutritionists, nutritional therapists, pharmacists, food service dieticians, public health nutritionists, sports nutritionists, and sports dietitians.

6.4   Digital Health Interventions

The digital health interventions category, with the proximal related term digital therapeutics, was defined in 2018 by the Digital Therapeutics Alliance as:

> *Digital therapeutics deliver evidence-based therapeutic interventions to patients that are driven by high quality software programs to prevent, manage, or treat a medical disorder or disease. They are used independently or in concert with medications, devices, or other therapies to optimize patient care and health outcomes. Digital therapeutics products incorporate advanced technology best practices relating to design, clinical validation, usability, and data security. They are reviewed and cleared or approved by regulatory bodies as required to support product claims regarding risk, efficacy, and intended use. Digital therapeutics empower patients, healthcare providers, and payers with intelligent and accessible tools for addressing a wide range of conditions through high quality, safe, and effective data-driven interventions.*

The first subcategory is m-health, with proximal related terms such as eHealth Devices, mobile health applications, or health apps. In 2020, the WHO defined m-health as:

> *Medical and public health practice supported by mobile devices, such as mobile phones, patient monitoring devices, personal digital assistances and other wireless devices* (WHO 2020d).

Professionals are psychotherapists, clinical psychologists, psychiatrists, licensed social workers, licensed professional counselors, and nurses.

The second subcategory is healthcare videogames with proximal related terms such as serious games, game-based digital interventions, digital therapeutic using video games, digital mental health games, or therapeutic games.

> Adapted from Saadatfard and Årsand in 2016, these interventions are *"used to combine novel interfaces (e.g. virtual reality headsets, mobile devices and wearable sensors) with a wide range of pedagogic approaches to deliver high-fidelity multimedia content for educational purposes. The following are the main areas for using serious games in healthcare. (1) Rehabilitation: The main aim of these games is to improve the cognitive and motor skills of patients during the rehabilitation process by making the exercises easier and more fun compared with traditional methods… (2) Health promotion and education: Depending on*

the target population, these games focus on aspects such as raising awareness, diet, exercise, hygiene and social abilities… (3) Distracting patients during painful medical procedures: The immersive characteristic of video games and virtual reality have been shown to be effective in focusing a patient's attention away from the pain caused by their treatment.

Professionals are psychotherapists, clinical psychologists, psychiatrists, licensed social workers, licensed professional counselors, and nurses.

The third subcategory is virtual reality therapies with proximal related terms such as augmented reality therapies or cybertherapies. Ventura et al. (2018) defined virtual reality therapies as follows:

The applied approach for the use of virtual reality (VR) and augmented reality (AR) on clinical and health psychology has grown exponentially. These technologies have been used to treat several mental disorders, for example, phobias, stress-related disorders, depression, eating disorders, and chronic pain. The importance of VR/AR for the mental health field comes from three main concepts: (1) VR/AR as an imaginal technology, people can feel "as if they are" in a reality that does not exist in external world; (2) VR/AR as an embodied technology, the experience to feel user's body inside the virtual environment; and (3) VR/AR as connectivity technology, the "end of geography".

Professionals are psychotherapists, clinical psychologists, and psychiatrists.

6.5   Elemental Health Interventions

The elemental health interventions category, with proximal related terms such as other non-pharmacological interventions or elemental resource health interventions, is defined as:

An elemental health intervention is an individual or group, non-invasive, described, and comprehensive program using an elementary resource for improving significantly health markers with a patently impact on health optimization, prevention or care.

The first subcategory is minerals with proximal related terms such as mineral methods, lithotherapies, evidence-based mineral methods, and evidence-based approach to minerals. A definition inspired from the NIH in 2020 states:

Minerals are essential substances that our bodies need to develop and function normally… A number of minerals are essential for health: calcium, phosphorus, potassium, sodium, chloride, magnesium, iron, zinc, iodine, sulfur, cobalt, copper, fluoride, manganese, and selenium.

Professionals are pharmacists, dieticians, and nutritionists.

The second subcategory is mycologycals with proximal related terms such as mycotherapy methods, mushroom therapies, mycotherapies, mushroom preparations, medicinal mushrooms, biological active mushrooms, evidence-based mushroom methods, or evidence-based approach to mushrooms. Popovic et al. (2013) defined mycologycals as:

Use of extracts and compounds obtained from mushrooms as medicines or health-promoting agents.

Professionals are pharmacists, dieticians, and nutritionists.

The third subcategory is botanicals with proximal related terms such as phyto-therapies, plant therapies, botanical preparations, botanical methods, herbal medi-cines, herbal supplements, herbal products, botanical products, phytomedicines, science-based herbalism, herbal remedies, herbals, plant-based therapies, botanical healthcare products, traditional herbal medicines, or herbal preparations. In 2004, the WHO stated that:

> Herbal medicines include herbs, herbal materials, herbal preparations and finished herbal products that contain as active ingredients parts of plants, other plant materials or combi-nations thereof. In some countries herbal medicines may contain, by tradition, natural organic or inorganic active ingredients that are not of plant origin (e.g. animal and mineral materials) (WHO 2004, p.6).

Professionals are pharmacists, dieticians, and nutritionists.

The fourth subcategory is electromagnetics with proximal related terms such as wave therapies, electromagnetic care, electromagnetic therapies, hyperthermia ther-apies, hypothermia therapies, plasmapheresis, or light therapies. Vadala et al. (2015) gave the following definition:

> Electromagnetic fields can be divided into two categories: static and time-varying. Electromagnetic therapy falls into two categories: (1) hospital use which includes TMS, repetitive transcranial magnetic stimulation (rTMS) and high-frequency TMS and (2) home use including PEMF therapy.

An example is therapeutic ultrasound programs. Professionals are mainly physiotherapists.

The fifth subcategory is cosmeceuticals with proximal related terms such as cos-metic therapies, cosmetic care, or socio-esthetical therapies. In 2011, Martin and Glazer defined cosmeceuticals as:

> A new category of products placed between cosmetics and pharmaceuticals that are intended for the enhancement of both the health and beauty of skin.

Professionals are socio-estheticians, estheticians, and pharmacists.

7 A Missing Link

7.1 Scientifically, a Space between Biomedical Treatments and Public Health Campaigns

Over the course of the last century, health was thought to be divided into two sepa-rated fields: biomedical treatments and health promotion. Biomedical treatments are extremely controlled, both before obtaining market access and being highly moni-tored after; their regulation procedures can be supranational (e.g., Europe). In health promotion, health institutes promoted education and awareness campaigns with attractive visual supports and intelligible recommendations (e.g., WHO campaigns about smoking). Countries replicated and culturally adapted the messages and content.

Dry January (UK 2013)

"Dry January started in 2013 with 4000 people. It's come a long way since then, with over 100,000 signing up and four million taking part in 2020. A month off is a great chance to get us all thinking about our drinking, so we can make healthier and happier decisions when it comes to alcohol year-round. Dry January is an opportunity to make not drinking, whether for an evening, a month or longer, feel more normal. Because many of us feel under pressure to drink, or to drink more than we want to, and we think it would be better if everyone had more choice."

World No Tobacco Day (WHO 2019a, b)

"Every year, on 31 May, the World Health Organization (WHO) and global partners celebrate World No Tobacco Day. The annual campaign is an opportunity to raise awareness on the harmful and deadly effects of tobacco use and second-hand smoke exposure, and to discourage the use of tobacco in any form. The focus of World No Tobacco Day 2019 is on "tobacco and lung health." The campaign will increase awareness on the negative impact that tobacco has on people's lung health, from cancer to chronic respiratory disease, the fundamental role lungs play for the health and well-being of all people. The campaign also serves as a call to action, advocating for effective policies to reduce tobacco consumption and engaging stakeholders across multiple sectors in the fight for tobacco control."

Your Healthiest Self: Wellness Toolkits (NIH 2020)

"Each person's "healthiest self" is different. We have different bodies, minds, living situations, and people influencing our lives. Each area can impact your overall health. This means we each have a unique set of health needs. Use our wellness toolkits to find ways to improve your well-being in any area you'd like.

Your Surroundings.

Learn how to make your environment safer and limit your exposure to potentially harmful substances to stay healthier.

Your Feelings.

Find out how to successfully handle life's stresses and adapt to change and difficult times.

Your Body.

Discover which physical health habits can help decrease your stress, lower your risk of disease, and increase your energy.

Your Relationships.

Learn how to create positive social habits that help you stay healthier mentally and physically.

Your Disease Defense.

Find out what steps you can take to protect your health and prevent diseases."
National Wear Red Day®: The First Friday in February (NHLBI 2020)
"Heart disease is largely preventable. Here's what you can do now to reduce your risk:

– Don't smoke,
– Eat for heart health,
– Aim for 30 min of physical activity at least 5 days a week,
– Ask your doctor to check your blood pressure, cholesterol, and blood glucose."

A missing link exists between biomedical products that ask patients to be passively treated and public health messages informing users without activating them to make significant and sustainable changes in their health behaviors. The space between them grows nowadays, and NPIs offer the bridge between them since these methods require users' active participation.

7.2 Legally, a Space between Biomedical Treatments and Consumer Products

It is interesting to note that in the current climate there is an increasing number of researchers working for the creation of a legal domain between biomedical technologies and consumer products. Biomedical technologies make spectacular efforts to keep patients alive with emergency and invasive solutions, largely used in hospitals. The gold standard is survival. Being qualified as a drug or medical device by authorities requires selling an argument and providing a concurrence restriction strategy. The challenges for biomedical companies are major, as are their investments. In the early part of the twentieth century, the U.S. FDA was given the responsibility for ensuring both the safety and efficacy of drugs prior to marketing. Amendments to the Federal Food, Drug, and Cosmetic Act in 1976 expanded the agency's role to oversee safety in the development of medical devices (Van Norman 2016). The approval of a new drug takes an average of 12 years whereas the approval of a new medical device from concept to market takes an average of 3–7 years (Fargen et al. 2013). Moving a medical device or drug from conception to market, as well as its post-market surveillance, is also expensive and uncertain.

On the other side, consumer products have to answer to basic needs and elicit the funds of clustered clients. Their access to market is direct, requiring only the respect of manufacturing and safety norms.

At the same time, new requests are coming from citizens concerned about a world facing ecological, sanitary, and demographic changes. These citizens are call-

Table 1.8 NPIs, a space between biomedical treatments and consumer products

Biomedical technologies	NPIs	Consumer products
Neurosurgeries Neurostimulations Radiation therapies Drugs	Animal-assisted therapies Behavioral change programs Education programs Psychotherapies	Entertainments
Biomedical devices Implantable tools Biomedical materials Orthosis/prosthesis Surgeries	Balneological therapies Exercise programs Manual therapies: Physiotherapies	Home materials Sport materials Vehicle Work products
Artificial nutrition Drugs	Diets Enriched/fortified foods Food supplements Novel foods Nutritional therapies Nutraceuticals Healthy eating pattern	Food products Restaurant
Implantable biotechnologies Medical devices	mHealth Serious games Virtual reality	Entertainment products Videogames
Drugs Medicines	Mushroom preparations Botanicals Mineral solutions	Food products
Plastic surgeries Skin injections Drugs	Cosmeceuticals	Cosmetics

ing for the health that was conceived of by the WHO in 1946 as real well-being combining longevity, autonomy, and health-related quality of life. They want to contribute to and be empowered in their own care and prevention. Thus, they want efficient and safe solutions to live better. Many solutions have emerged in the space between biomedical treatments and consumer products listed in Table 1.8. "*Cosmeceuticals*" have undoubtedly taken over the personal care industry across the globe. Despite the prevalent confusion about its definition and scope, it would not be an exaggeration to state that almost 30–40% of any dermatologist's prescription count across the world consists of a cosmeceutical (Pandey et al. 2020). The term was coined in 1984 by Dr. Albert Kligman of the University of Pennsylvania describing a hybrid category of products midway on the spectrum of cosmetics and pharmaceuticals. Their indications for use include antiaging in general; the treatment of photomelanosis and photo tanning; the treatment of pigmentation-related disorders like melasma or freckles rhytide reduction; anti-inflammatory use; fat loss; hair growth; the prevention of hair loss; and the maintenance of skin tone, skin clarity, and complexion. Physicians educated in these cosmeceuticals can serve their patients by (1) managing expectations for realistic, evidence-based effects, and (2) warning against and monitoring for potential side effects (Martin and Glaser 2011). "*Nutraceuticals*" is another example. These products may be used to improve health, delay the aging process, prevent chronic diseases, increase life expectancy, or support the structure or function of the body. Nowadays, nutraceuticals have received considerable interest due to potential nutritional, safety, and therapeutic effects (Nasri et al. 2014).

Cosmeceuticals
"Currently, cosmeceuticals are a segregated subclass within the domain of a cosmetic or drug. In Europe and Japan, cosmeceuticals are a subclass of cosmetics; however, in the US, cosmeceuticals can only be considered as a subclass of drugs" (Pandey et al. 2020).

The creation of a specific domain of NPIs will be a selling argument for professionals and a guarantee of quality for citizens. As such, the need for a national or supranational label is paramount. This is a domain claiming health-added value, evidence-based practice, scalable methods, visibility, credibility, prescription/recommendation/counselling from health professionals, traceability, and responsibility for its professionals. Furthermore, this domain can become an antechamber for some drugs or medical devices. It is an area that encourages professionals to continuously improve the quality of their protocols and the delivery of their expertise.

7.3   An Ideal World

Ideally, a doctor should be able to prescribe NPIs. Depending on diagnosis, she or he would indicate the most relevant NPIs for each patient. She or he would explain their mechanisms of action, their benefits, their risks, and their implementation. She or he would anticipate interactions with other authorized biomedical treatments. She or he would suggest a dosage to the patient by personalizing it in accordance with that patient's preferences and lifestyle. She or he would share the decision with the patient. She or he would advise the patient on what to do. She or he would refer to a network of qualified and certified professionals located near the patient's places of life. She or he would follow the progress of that patient at a distance. She or he would set an appointment schedule that is in line with the patient's therapeutic objectives. She or he would report to the competent authorities on adverse effects and harmful interactions with other therapies.

8 Conclusion

An intermediate space between biomedical treatments and health promotion has been under construction by science since the beginning of the century. Its advent is inexorable throughout the world, even if its economic and regulatory recognition has yet to be established. The first step requires abandoning nebulous terms qualifying protocols as complementary and alternative medicine or natural medicine. These terms are too ambiguous, including a pell-mell that mixes cross-categorically across appellations, diagnostic processes, approaches, traditional practices, and sectarian practices. Non-pharmacological interventions (NPIs) or human and natural health methods (HUNAMs) constitute a whole domain of evidence-based and scalable methods dedicated to health, autonomy, longevity, and health-related quality of life.

Key Points
The term NPIs needs to be used instead of complementary or alternative medicine or their equivalents because NPIs:

- Respect the diagnosis of modern Western medicine.
- Are resolutely based on science and a quality approach.
- Are transparently described in a manual.
- Are noninvasive.
- Have health benefits and risks documented in rigorous studies published in independent peer-reviewed scientific journals.
- Are implemented by specifically trained professionals.
- Are integrated into most national health strategies in the world.
- Are increasingly reimbursed by health insurance, insurance, and provident fund.

  Each NPI has:

- A primary health goal.
- A presentation manual for the user.
- An implementation manual with full descriptions for professionals.
- References of Published Studies in Peer-Reviewed Scientific Journals Showing Benefits and Risks
- A guarantee of continuous improvement process (e.g., study in progress, scientific conference presentation).
- A specifically trained professional.

References

American Art Therapy Association. (2013). https://arttherapy.org/about/
American Horticultural Therapy Association. (1997). https://www.ahta.org/ahta-definitions-and-positions
American Psychiatric Association. (2019). https://www.psychiatry.org/patients-families/psychotherapy
American Veterinary Medical Association. (2020). https://www.avma.org/policies/animal-assisted-interventions-definitions
Apelian, N., & Davis, C. (2018). *The lost book of herbal remedies*. Austin: Capital Printing.
Balneology Association of North America. (2020). https://balneology.org/introduction-to-balneology/
Boutron, I., Ravaud, P., & Moher, D. (2012). *Randomized clinical trials of non pharmacological treatments*. Bacon Raton: CRC Press Taylor and Francis.
Casarin, A. I., Tangkiatkumjai, M., & Walker, D. M. (2019). An overview of complementary and alternative medicine. In Information Resources Management Association (Ed.), *Complementary and alternative medicine: Breakthroughs in research and practice* (pp. 1–16). Hershey: IGI Global.
Centre de Recherche et d'Information des Organisations de Consommateurs. (2012). *Les Médecines alternatives*. Brussels: CRIOC.
Chartered Society of Physiotherapy. (2020). https://www.csp.org.uk/careers-jobs/what-physiotherapy
Clar, C., Tsertsvadze, A., Court, R., Lewando Hundt, G., Clarke, A., & Sutcliffe, P. (2014). Clinical effectiveness of manual therapy for the management of musculoskeletal and non-musculoskeletal conditions: Systematic review and update of UK evidence report. *Chiropractic and Manual Therapies, 22*(1), 12. https://doi.org/10.1186/2045-709X-22-12.
ClinicalTrials. (2020). https://clinicaltrials.gov/ct2/about-studies/learn
Digital Therapeutics Alliance. (2018). https://www.who.int/goe/publications/goe_mhealth_web.pdf
Dry January. (2013). https://alcoholchange.org.uk/get-involved/campaigns/dry-january/about-dry-january/the-dry-january-story
Fargen, K. M., Frei, D., Fiorella, D., McDougall, C. G., Myers, P. M., Hirsch, J. A., & Mocco, J. (2013). The FDA approval process for medical devices: An inherently flawed system or a valuable pathway for innovation? *Journal of Neurointerventional Surgery, 5*, 269–275. https://doi.org/10.1136/neurintsurg-2012-010400.
French National Authority for Health. (2011). *Développement de la prescription de thérapeutiques non médicamenteuses validées*. Paris: HAS.
French Senate. (2013). *Rapport 480 au nom de la commission d'enquête sur l'influence des mouvements à caractère sectaire dans le domaine de la santé*. Paris: Senate.
Glasziou, P., Meats, E., Heneghan, C., & Shepperd, S. (2008). What is missing from descriptions of treatment in trials and reviews? *British Medical Journal, 336*(7659), 1472–1474. https://doi.org/10.1080/01421590802572791.
Hoffmann, T. C., Erueti, C., & Glasziou, P. P. (2013). Poor description of non-pharmacological interventions: Analysis of consecutive sample of randomised trials. *British Medical Journal, 347*, f3755. https://doi.org/10.1136/bmj.f3755.
Institute of Health Sciences. (2020). https://instituteofhealthsciences.com/what-is-nutritional-therapy/
Kenny, M., Luck, P., & Koerbel, L. (2020). Tending the field of mindfulness-based programs: The development of international integrity guidelines for teachers and teacher training. *Global Advance in Health Medicine, 9*. https://doi.org/10.1177/2164956120923975.
Khalil, M. K. M., Al-Eidi, S., Al-Qaed, M., & AlSanad, S. (2018). The future of integrative health and medicine in Saudi Arabia. *Integrative Medicine Research, 7*, 316–321. https://doi.org/10.1016/j.imr.2018.06.004.

Li, Q. (2010). Effect of forest bathing trips on human immune function. *Environmental Health and Preventive Medicine, 15*(1), 9–17. https://doi.org/10.1007/s12199-008-0068-3.

Lognos, B., Carbonnel, F., Boulze-Launay, I., Bringay, S., Guerdoux-Ninot, E., Mollevi, C., Senesse, P., & Ninot, G. (2019). Complementary and alternative medicine in patients with breast cancer: An exploratory study of social network forums data. *Journal of Medical Internet Research Cancer, 5*(2), e12536. https://doi.org/10.2196/12536.

Martin, K. I., & Glaser, D. A. (2011). Cosmeceuticals: The new medicine of beauty. *Missouri Medicine, 108*(1), 60–63.

Mayo Clinic. (2017). *Mayo clinic book of home remedies* (2nd ed.). New York: Time Book.

Nasri, H., Baradaran, A., Shirzad, H., & Rafieian-Kopaei, M. (2014). New concepts in nutraceuticals as alternative for pharmaceuticals. *International Journal of Preventive Medicine, 5*(12), 1487–1499.

National Cancer Institute. (2020). https://www.cancer.gov/publications/dictionaries/cancer-terms/def/mind-body-practice

National Health Service. (2020). https://www.nhs.uk/conditions/complementary-and-alternative-medicine/

National Wear Red Day. (2020). https://www.nhlbi.nih.gov/health-topics/education-and-awareness/heart-month/wear-red-day

NIH. (2020). https://www.nccih.nih.gov/health/vitamins-and-minerals

Ninot, G. (2013). *Démontrer l'efficacité des interventions non médicamenteuses: Question de points de vue.* Montpellier: Presses Universitaires de la Méditerranée.

Ninot, G. (2019). *Guide professionnel des interventions non médicamenteuses (INM).* Paris: Dunod.

Pandey, A., Jatana, G. K., & Sonthalia, S. (2020). *Cosmeceuticals* (p. 2020). Treasure Island: StatPearls Publishing.

Plateforme CEPS. (2017). https://plateformeceps.www.univ-montp3.fr/fr/nos-services/classificationinm

Popović, V., Živković, J., Davidović, S., Stevanović, M., & Stojković, D. (2013). Mycotherapy of cancer: An update on cytotoxic and antitumor activities of mushrooms, bioactive principles and molecular mechanisms of their action. *Current Topics in Medicinal Chemistry, 13*(21), 2791–2806. https://doi.org/10.2174/15680266113136660198.

Rakel, D. (2018). *Integrative medicine* (4th ed.). Philadelphia: Elsevier.

Rakel, D., & Weil, A. (2018). Philosophy of integrative medicine. In D. Rakel (Ed.), *Integrative medicine* (4th ed.). Philadelphia: Elsevier.

Saadatfard, O., Årsand, E. (2016). https://ehealthresearch.no/files/documents/Faktaark/Fact-sheet-2016-11-Serious-Games-in-Healthcare.pdf

Therapeutic Goods Administration. (2020). https://www.tga.gov.au/overview-regulation-complementary-medicines-australia

Vadalà, M., Vallelunga, A., Palmieri, L., Palmieri, B., Morales-Medina, J. C., & Iannitti, T. (2015). Mechanisms and therapeutic applications of electromagnetic therapy in Parkinson's disease. *Behavioral and Brain Function, 11*, 26. https://doi.org/10.1186/s12993-015-0070-z.

Van Norman, G. A. (2016). Drugs, devices, and the FDA: Part 2: An overview of approval processes: FDA approval of medical devices. *JACC: Basic to Translational Science, 1*(4), 277–287. https://doi.org/10.1016/j.jacbts.2016.03.009.

Ventura, S., Baños, R. M., & Botella, C. (2018). Virtual and augmented reality: New frontiers for clinical psychology. In N. Mohamudally (Ed.), *State of the art virtual reality and augmented reality knowhow.* Intechopen.

WHO. (2004). *WHO guidelines on safety monitoring of herbal medicines in pharmacovigilance systems.* Geneva: WHO.

WHO. (2013). *WHO traditional medicine strategy: 2014–2023.* Geneva: WHO.

WHO. (2017a). https://www.who.int/mental_health/management/scalable_psychological_interventions/en/

WHO. (2017b). https://www.who.int/features/factfiles/nutrition/en/

WHO. (2019a). *WHO global report on traditional and complementary medicine 2019*. Geneva: WHO.
WHO. (2019b). https://www.who.int/classifications/ichi/en/
WHO. (2020a). https://www.who.int/health-topics/traditional-complementary-and-integrative-medicine
WHO. (2020b). https://www.who.int/topics/health_education/en/
WHO. (2020c). www.who.int/dietphysicalactivity/pa/en/
WHO. (2020d). https://www.who.int/goe/publications/goe_mhealth_web.pdf
Wieland, S. L., Manheimer, E., & Berman, B. M. (2011). Development and classification of an operational definition of complementary and alternative medicine for the Cochrane collaboration. *Alternative Therapy and Health Medicine, 17*(2), 50–59.
Wikipedia. (2020). https://en.wikipedia.org/wiki/Alternative_medicine
Willett, W., & Skerrett, P. J. (2011). *Eat, drink, and be healthy: The Harvard Medical School guide to healthy eating*. New York: Free Press.
World Cancer Research Fund. (2018). https://www.wcrf.org/dietandcancer/recommendations/dont-rely-supplements
World No Tobacco Day. (2019). https://www.who.int/news-room/events/detail/2019/05/31/default-calendar/world-no-tobacco-day
Your Healthiest Self: Wellness Toolkits. (2020). https://www.nih.gov/health-information/your-healthiest-self-wellness-toolkits

Chapter 2
The Reasons for the Success of Non-pharmacological Interventions

1 Introduction

Non-pharmacological interventions (NPIs) have become essential solutions for better living, human performance, disease prevention, behavioral change, longevity, and improved quality of life. They have also become a necessary complement of biomedical treatments, enhancing their effects, limiting their side effects, and improving patient compliance. NPIs constitute a promising ecosystem in terms of innovation, market development, and as a contributor to integrative health. Their success is such that not a day goes by without a study publication in a scientific journal, a press article, hundreds of posts on social networks, or a television report about them. What is more, this phenomenon can be observed in every country. This chapter explains the 10 main reasons for this success.

2 The Omics Revolution

2.1 Genetics

Once understood on the basis of the Latin root *sano, sanare* – to make healthy, to repair – health represented an ordered and stable set of cells. This conception supported the hope that any organ could be reconditioned, strengthened, transformed, or replaced. As a result, this notion opened the door to biotechnological prowess, from the macroscopic level with surgeries and biological/mechanical transplants to the microscopic level with targeted drugs and radiotherapy.

Nothing seemed to stop this infinitely small analytical approach. It followed the path traced by Louis Pasteur and Claude Bernard as well as the fundamental philosophical thought of Aristotle and René Descartes: identifying a diseased organ or

© The Author(s), under exclusive license to Springer Nature Switzerland AG 2021
G. Ninot, *Non-Pharmacological Interventions*,
https://doi.org/10.1007/978-3-030-60971-9_2

failing function, experimentally establishing its biological cause-and-effect relationship, and finding a solution that acts on this relationship in order to systematically, if not always, cure the disease.

The Attraction of Deterministic Medicine According to Professor Claude Bernard in 1865:

"We must recognize in all science two classes of phenomena, some whose cause is currently determined, others whose cause is still undetermined. For all the phenomena whose cause is determined, statistics have no place; their use would even be absurd. Thus, as soon as the circumstances of the experiment are well established, we can no longer generate statistics: we will not, for example, collect the cases to know how often water is formed from oxygen and hydrogen; to know how many times cutting the sciatic nerve will cause the paralysis of the muscles to which it is connected. These effects will always happen without exception, because the cause of the phenomenon is exactly determined. It is therefore only when a phenomenon contains as yet undetermined conditions that statistics can be generated; but what you need to know is that you only use statistics because you can't do anything else; because statistics, according to me, can never give scientific truth and therefore, cannot constitute a definitive scientific method" (Claude Bernard 1865, Chapter II; translated by the author).

This experimental and hypothetical deductive-reductive approach has divided medicine into organ specialties (e.g., cardiology, pneumology, neurology, dermatology) and functional specialties (e.g., endocrinology).

Deterministic and reductionist medicine experienced its heyday in the twentieth century, notably in the fields of infectious diseases, by eradicating pandemic diseases (e.g., malaria, plague) and acute diseases (e.g., syphilis), and by introducing mass treatments (e.g., penicillin). The successes of a single treatment that worked for any "average individual" gave hope and confidence that we would eradicate all diseases. Gene therapy embodies the ultimate accomplishment of this logic – any faulty gene can be replaced. Ever since "Dolly" the sheep demonstrated that cloning is now under our control, we believe that nothing prevents genetic repair. Genetics is no longer a promise, but rather it seems to be a reality. The population only had to be patient, even as a patient.

In short, this approach that is embodied by genetics and biocompatible nanotechnologies is stirring up all fantasies, gathering all promising innovations, concentrating all research efforts and all investments.

This begs the question, if robots need to be repaired, do not humans prefer to be cared?

2.2 Epigenetics

The beginning of the twentieth century brought hope for the triumph of biochemistry and biotechnology over all diseases. The end of the twentieth century maintained this promising hope with gene therapies and biotech replacements. Unfortunately, the twenty-first century is telling us another story about how we should consider health. Recalling that this conception comes from the Latin verb *saluto, salutare,* meaning to keep safe or to preserve over time, health is a precious and fragile good. In Greek mythology, health obeys the precepts of the goddess *Hygie* (*Hygiee*), goddess of health, cleanliness, and hygiene. She was the daughter of one of the most popular divinities, *Asclepius,* the god of medicine, and the nymph *Epione,* the daughter of the King of Cos. Contrary to the popular belief that the body is simply an inert mass of genes that is inherited from an individual's parents, recent studies support the ancient myth that our genome is the product of complex and constantly evolving interactions between the organism and its environment. Researchers call this interaction epigenetics (Carey 2012).

Defining Epigenetics

Epigenetics is the study of the constant evolution of the genome as a function of the environment, behavior, social context and cultural development. This evolution does not occur over several generations; it is not on the scale of a generation, rather within a few months in the same individual (Carey 2012).

Derived from the Greek *epigenesis,* epigenetics tries to elucidate the molecular basis of Conrad Waddington's observations in which environmental stress caused the genetic assimilation of certain phenotypic characteristics in Drosophila fruit flies. Researchers improved our understanding of the epigenetic mechanisms related to these types of changes. The genome refers to the entire set of genetic information as a nucleotide sequence within the DNA, whereas the epigenome refers to complex modifications within genomic DNA. Epigenetics not only considers genomic constitution, but also integrates the social and natural environment, the influence of everyday routines, dietary habits, and stresses to biological systems. The epigenome integrates information that is encoded in the genome with molecular and chemical cues of cellular, extracellular, and environmental origin to define the functional identity of each cell type during development or disease. These stimuli-initiated modulations of the epigenome contribute to embryo development, cell differentiation, and responses to exogenous signals. Thus, in contrast to the consistency of the genome, plasticity in the epigenome is characterized by dynamic and flexible responses to intracellular and extracellular stimuli including those from the environment. Modifications in the genome regulate numerous cellular activities, and disruption of these activities may cause abnormal expression or silencing of genes. DNA methylation is one of the most broadly studied and well-characterized epigenetic modifications. This research suggests that DNA methylation may be important in long-term memory function. Other major modifications include chromatin remodeling, histone modifications, and non-coding RNA mechanisms. Unlike most genetic defects, epigenetic defects are reversible. This reversibility is an important facet of the epigenetic contribution to diseases and makes such diseases amenable to treatments.

Several studies of epigenetics show how different lifestyles can alter marks on top of DNA and play a role in determining health outcomes. For example, researchers have found that a ketogenic diet – consuming high amounts of fat, adequate protein, and low carbohydrates – increases an epigenetic agent naturally produced by the body. Diet has also been shown to modify epigenetic tags in significant ways. Several studies have found that certain compounds within the foods we consume can protect against cancer by adjusting methyl marks on oncogenes or tumor suppressor genes. For example, a study has showed the ability of a nutrient in garlic – an organosulfur compound called S-allylcysteine – to reduce ovarian cancer cell growth and to have an epigenetic mechanism of action in the form of DNA methylation through DNA methyltransferases 1 (Xu et al. 2018). An epigenetic diet may guide people toward an optimal food regimen as more scientific studies reveal the underlying mechanisms and impact that different foods have on the epigenome and health.

Exercise Program: An Epigenetic Remedy?

A Swedish study by Lindholm et al. (2014) assessed 12 men and 11 women who were healthy but untrained. These 23 participants, who were 27 years old on average, followed an endurance training program that was supervised by a professional. Participants used a machine allowing them to do flexions and extensions with one leg. This program lasted 3 months, at the rate of 4 sessions for 45 minutes per week. The originality of this study was that the training program involved only one leg, which allowed the other to serve as a control comparison. A sample from the trained leg was drawn and muscle performance was measured before and after training. Markers of muscle metabolism, including the methylation state of 480,000 genome sites and the activity of 2000 genes, were analyzed following a biopsy of the thigh muscle, *vastus lateralis*. The results showed a strong relationship between epigenetic methylation and changes in the activity of 4000 genes. Genes associated with genomic regions in which methylation levels increase are involved in the adaptation of skeletal muscles and carbohydrate metabolism, while a decrease appears in regions associated with inflammation. Researchers find that the majority of epigenetic changes occur in the regulatory regions of the human genome – the enhancers. Researchers have also noted epigenetic differences in the muscles of women and men. This study concluded that a simple endurance training of 3 h per week for 3 months impacts thousands of DNA methylation sites and genes associated with improved muscle functioning and health (metabolism and inflammation).

Moreover, several studies of epigenetics have also shown several ways in which different environmental exposures can alter markers on top of DNA and play a role in determining health outcomes. For example, researchers have found that air pollution can alter methyl tags on DNA and increase one's risk for neurodegenerative disease. Interestingly, B vitamins may protect against the harmful epigenetic effects of pollution and may be able to combat the harmful effects that particular types of matter have on the body.

Epigenetics overturns fixed conceptions of genetic heritage, of organs that wear down with age, of a brain promised to degenerate, and of a personality that is frozen after birth. Seventy percent of our health depends on our environment; so, we can influence it.

2.3 Personomics

It was a great disappointment to the proponents of hereditary health when it was found that the brain of one of the greatest geniuses of the twentieth century – Albert Einstein – was perfectly anatomically normal! Nevertheless, neurosurgeons have observed the tremendous adaptive capacity of the brain through brain plasticity. The discovery of stem cells, including stem cells in the brain, has upset deterministic conceptions of fixedness from birth. Psychological resilience also shows to what extent a life event, like an illness, can transform a personality that is supposed to be immutable. Athletes prove how much training pushes their limits.

Who is not impressed by Mike Horn's explorations at the end of the world? Who is impassive to the prowess of the ultra-trail runner Kylian Jornet? Who is not captivated by the stories of the deep-sea diver Jacques Mayol? Who hasn't heard of the centenary French cyclist Robert Marchand? Who is not amazed by the performances of tennis player Roger Federer, no. 1 at only 36 years old? Man has pushed his limits since the dawn of time. He has turned his adaptability into one of his cardinal virtues which has placed him at the top of the food chain.

The concept of personomics emphasizes components of individuality, such as the previously noted concepts of genetics and epigenetics, as being critical to patient care. According to Ziegelstein (2017), the concept restores a balance between prevention and therapy, between care and caring, between genetic heritage and lifestyle. Individuals need to be distinguished not only by their biological variability, but also by their personalities, health beliefs, social support networks, financial resources, and other unique life circumstances that have important effects on how and when a given health condition will manifest in that individual and how that condition will respond to treatment.

In this era of personomics, there is ample place for NPIs. They stimulate the internal resources of the body's biology as much as they modify our understanding of the environment and our relationships with other humans. They encourage people to look for solutions for self-healing and favor endomedicines.

Toward Personomics

"When medicine is informed solely by clinical practice guidelines, however, the patient is not treated as an individual, but rather a member of a group. Precision medicine, as defined herein, characterizes unique biological characteristics of the individual or of specimens obtained from an individual to tailor diagnostics and therapeutics to a specific patient. These unique biological characteristics are defined by the tools of precision medicine: genomics, proteomics, metabolomics, epigenomics, pharmacogenomics, and other "-omics." Personalized medicine, as defined herein, uses additional information about the individual derived from knowing the patient as a person. These unique personal characteristics are defined by tools known as personomics which takes into account an individual's personality, preferences, values, goals, health beliefs, social support network, financial resources, and unique life circumstances that affect how and when a given health condition will manifest in that person and how that condition will respond to treatment. In this paradigm, precision medicine may be considered a necessary step in the evolution of medical care to personalized medicine, with personomics as the missing link" (Ziegelstein 2017, p. 2).

They remind us that man is a unified whole – a brain and a body that interact with an environment. As such, acting on one component can have cascading benefits on other parts of the body. Thus, it remains yet to be discovered the best ways and for whom NPIs are really profitable.

The last century made it possible for us to identify targeted pathophysiological mechanisms and to discover mass treatments that are able to cure some organic diseases. While this approach brought hope to our attempt to discover curative treatments for every/disease, it is now clear that many so-called chronic, progressive, and multifactorial (genome, epigenome, behavior, environment, pathogen) diseases escape it: diabetes, obesity, cancers, cardiovascular diseases, respiratory diseases, joint diseases, neurodegenerative diseases, anxiety disorders, depressions, and fibromyalgia. These diseases become more complex over time, accumulating with aging and precariousness in all its forms (economic, social, and environmental). As a result, these complex diseases do not correspond to Claude Bernard's linear, universal, and restricted explanations.

3   Health Impacts of Behaviors

A famous precept of Hippocrates (460–377 BC) underscores how important physical activity and diet are for health. Today, studies confirm how much skillfully dosed physical activities and adapted diets support healthy aging. Others show that these solutions are also beneficial for people with disease. These behaviors carry the seeds of effective solutions for longevity.

Hippocratic Precept
"If we could give each individual the right amount of food and exercise, not too little and not too much, we would have found the safest way to health."

However, other behaviors that some epicurean artists and intellectuals – like Oscar Wilde – have taken a mischievous pleasure in promoting have the opposite effect. Some unhealthy lifestyle factors counterbalance gains in life expectancy; in particular, the increasing obesity epidemic and decreasing levels of physical activity. Research has shown the deleterious effects of behaviors linked to alcohol, tobacco, poor diet (e.g., too fatty, too salty, too sweet, too abundant, too limited, too unbalanced) and physical inactivity, implicating these behaviors in years of life lost, organ damage, and increased pain at the end of life. Other behaviors that can be included are drug non-compliance, insufficient body hygiene (especially tooth and hand hygiene), psychotropic abuse, gaming addictions, and unprotected sexual activity.

Radical Epicurean Precept
"To get back my youth I would do anything in the world, except take exercise, get up early, or be respectable" (Oscar Wilde, The Picture of Dorian Gray, 1891).

A healthy lifestyle has been associated with an estimated increased life expectancy of 8.3 years for women and 10.3 years for men in Japan (Tamakoshi et al. 2009); 13.9 years for women and 17.0 years for men in Germany (Li et al. 2014); 17.9 years in Canada (Manuel et al. 2016); and of engendering a 14 years' difference in chronological age in the United Kingdom (Khaw et al. 2008). Data from three European cohorts in Denmark, Germany, and Norway suggested that men and women of 50 years of age with favorable lifestyles would live 7.4 to 15.7 years longer than those with an unfavorable lifestyle (O'Doherty et al. 2016). A U.S. cohort study with 123,219 adults (health professionals) showed that adherence to five low-risk, lifestyle-related factors (never smoking, maintaining a healthy weight with a body mass index between 18.5 and 25 kg/m^2, practicing regular physical activity defined as at least 3.5 h per week of moderate to vigorous intensity activity, following a healthy diet, and moderating alcohol consumption to an intake of 5 to 15 g/d for females and 5 to 30 g/d for males) could prolong life expectancy, at 50 years of age, by 14.0 years for females and 12.2 years for males compared with individuals who adopted zero low-risk lifestyle factors (Li et al. 2018). The combination of these five factors is powerful with a cumulative effect: the larger the number of low-risk lifestyle factors, the longer the potential prolonged life expectancy, regardless of the combined factors. The life expectancy benefits of this lifestyle pattern are still observed after 70 years of age, even if the impact is obviously less. Each individual component of a healthy lifestyle has also showed a significant association with the risks of total mortality, cancer mortality, and cardiovascular disease mortality. A combination of five low-risk lifestyle factors was associated with a hazard ratio of 0.26 for all-cause mortality (95% CI, 0.22–0.31), 0.35 for cancer mortality (95% CI, 0.27–0.45), and 0.18 for cardiovascular disease mortality compared to participants with zero low-risk factors (95% CI, 0.12–0.26). The population-attributable risk of nonadherence to five low-risk lifestyle factors was 60.7% for all-cause mortality (95% CI, 53.6–66.7), 51.7% for cancer mortality (95% CI, 37.1–62.9%), and 71.7% for cardiovascular mortality (95% CI, 58.1–81.0). A similar association was observed between low-risk lifestyle factors and mortality before 75 years of age. Low-risk lifestyle factors were also associated with lower risk of cause-specific mortality in women and men as well (Li et al. 2018).

Health behaviors play a major role in the underlying causes of chronic noncommunicable diseases (NCDs). According to the 2010 global burden of disease study, 7 of the top 12 and 10 of the top 20 causes of all disability-adjusted life years lost across the world are due to poor health behaviors such as unhealthy dietary habits, low physical activity levels, and smoking (Lim et al. 2012). The fact that medication nonadherence was not considered as a potential cause of disability suggests that the global burden of unhealthy behaviors may be even greater than is currently recognized. Interestingly, though genetic factors may account for some variability in the development of cardiovascular diseases, systematic assessments have estimated that ">70% of total cardiovascular events, 80% of coronary heart disease events, and 90% of new cases of diabetes mellitus" can be attributed to a few basic health behaviors; that these same behaviors influence both established (e.g., dyslipidemia, hypertension, and diabetes mellitus) and novel (e.g., inflammation, endothelial function, thrombosis, and arrhythmia) cardiovascular disease risk factors; and that risk biomarkers and behaviors are similarly related to cardiovascular disease events (Mozaffarian et al. 2008).

Not only do poor health behaviors predict disease development and outcomes, but improving these behaviors (quality and sustainability) also has been shown to prevent and reduce the severity of chronic diseases. Even modest alterations in health behaviors have been shown to have positive effects on cardiovascular disease, cancer, respiratory disease, and neuropsychiatric disease risk (Bacon et al. 2015). For example, smoking prevention and cessation programs have had significant direct impacts on cancer, cardiovascular disease, and chronic obstructive pulmonary disease outcomes, and quitting smoking is associated with a 33% reduction in total mortality (Critchley and Capewell 2003).

Enhancing Sustainable Health Behavior Changes Is Better Than Trying to Treat Incurable Disease

Prevention is better than cure, and yet everything is done, especially in France, to encourage people to consult a doctor as late as possible. Biomedical therapies are reimbursed, while prevention actions are questioned and highly debated. In the twentieth century the curative approach prevailed over the preventive approach; now the twenty-first century is slated to rebalance health solutions. The national strategies, plans, and statements about cancer, neurodegenerative diseases, addictions, mental health conditions, chronic diseases, healthy aging, and active mobility that have been disseminated since 2000 encourage this movement. Academic reports and grants have accelerated innovations in health prevention strategies. Europe encourages each nation to pursue a coherent, concerted, and multichannel prevention policy (Article 168 of the Treaty on the Functioning of the European Union). The WHO, in particular, supports it through the promotion of training for health, educational, and social professionals (WHO 2006). This movement requires both the implementation of comprehensive policies favoring health promotion and concrete and targeted prevention actions. These prevention actions with targeted methods, content, and objectives are NPIs. In primary prevention, they prevent diseases. In secondary prevention, they delay the worsening of diseases. In tertiary prevention, they avoid complications from diseases.

Integrated as NPIs, *"behavioral interventions as those that require the active participation of a target group (e.g., the patient/individual, partner, health professional, a community/population, health care systems) in a program (e.g., a dose and an administration mode) delivered by a trained interventionist with the proximal or ultimate goal of changing health-related behavior (e.g., eating a healthier diet, smoking cessation, medication adherence, promotion of physical exercise, and self-management of chronic disease), all of which have been shown to be beneficial"* (Bacon et al. 2015). Examples of these benefits include reduced symptoms of chronic disease (e.g., dyspnea in chronic obstructive pulmonary disease), decreased side effects of treatments (e.g., fatigue in cancer), decreased risk of mental health and medical comorbidities (e.g., depression, obesity), improved quality of life, reduction in direct disease-care costs, and decreases in indirect expenditure (e.g., work absenteeism). In addition, there seem to be collateral benefits in that caregivers, families, and friends of those engaging in such interventions also reduce their risk of NCDs.

4   Adding Life to Years, Not Just Years to Life

The U.S. life expectancy at birth has doubled from 40 years in 1850 to 80 years in 2000. In 1950, approximately 8% of the population (12 million people) was over 65 years of age. By 2015, the percentage had almost doubled (15%, 45 million people). Americans reaching the age of 65 can expect, on average, to live an additional 19 years. Not only do people who are getting older want to extend their longevity, they also want to have a better quality of life. As such, the near future will strongly focus on ensuring that citizens experience a full life for as long as possible. The debate is centered on questions of whether these extra years are spent in a good or a poor quality of life. Healthcare indicators have traditionally included survival rates, standardized mortality ratios, and avoidable mortality. Now, health care is focusing more on patients' reported outcomes as a means of redirecting attention toward considering the impact that a condition or treatment has on the patient's emotional and physical functioning and lifestyle. Quality-of-life markers help to answer questions of whether treatments lead to a life worth living by providing a more patient-centered baseline against which the effects of interventions can be evaluated. This movement was initiated and has spread considerably among people suffering from chronic diseases. In the case of chronic conditions, in which there can only be a partial or temporary improvement of symptoms, the realistic goal of care is to provide a life that is as comfortable, functional, and satisfying as possible.

The World Health Organization (WHO) was among the first institutions to raise an alarm about the rapidly expanding pandemic in the world; that of noncommunicable diseases (NCDs), also called chronic diseases: *"Persistent communicable (HIV / AIDS) and non-communicable diseases (cardiovascular pathologies, cancers and diabetes), certain mental disorders (depression and schizophrenia), as well as permanent physical handicaps (amputations, blindness and joint diseases), they may appear different, they all fall into this category"* (WHO 2003, p.11). NCDs are responsible for 70% of the world's deaths (WHO 2013a) and health expenditures (WHO 2008).

> **Noncommunicable Disease Mortality in the World**
> Noncommunicable diseases kill more than 40 million people each year, accounting for 70% of the world's deaths. Forty-four percent of these deaths are caused by cardiovascular disease, 22% by cancer, 10% by respiratory disease, and 4% by diabetes. These deaths are premature, since 15 million people die between 30 and 69 years old (WHO 2013a).
>
> "NCDs currently cause more deaths than all other causes combined and NCD deaths are projected to increase from 38 million in 2012 to 52 million by 2030. Four major NCDs (cardiovascular diseases, cancer, chronic respiratory diseases, and diabetes) are responsible for 82% of NCD deaths" (WHO 2014, p. 8).

Table 2.1 Variation of years of life with a disability for main communicable and noncommunicable diseases (GBD 2017 Disease and Injury Incidence and Prevalence Collaborators 2018)

	Percentage change, 1990–2007	Percentage change, 2007–2017
Noncommunicable diseases		
Neoplasms (e.g., breast cancer, prostate cancer)	+59.3%	+40.6%
Hypertensive heart disease	+63.8%	+34.9%
Diabetes mellitus (e.g., type 2 diabetes)	+75.9%	+31.1%
Chronic respiratory diseases (e.g., asthma, COPD)	+21.9%	+22.8%
Musculoskeletal disorders (e.g., low back pain)	+38.4%	+19.9%
Neurological disorders (e.g., Alzheimer)	+35.1%	+17.8%
Substance use disorders (e.g., alcohol dependence)	+34.3%	+16.7%
Depressive disorders (e.g., dysthymia)	+33.4%	+14.3%
Mental disorders (e.g., anxiety disorders)	+31.6%	+13.5%
Skin and subcutaneous diseases (e.g., dermatitis)	+24.0%	+13.0%
Communicable diseases		
HIV/AIDS and sexually transmitted infections	+204.0%	−6.0%
Nutritional deficiencies (e.g., vitamin A deficiency)	−8.3%	−7.3%
Neglected tropical diseases and malaria (e.g., dengue)	+2.4%	−10.3%

Source: Adapted from GBD 2017 Disease and Injury Incidence and Prevalence Collaborators (2018), licensed under the terms of the Creative Commons Attribution License (https://creative-commons.org/licenses/by/4.0/). Copyright © 2018 The Author(s). Published by Elsevier Ltd.

Chronic diseases are characterized "*by the extent of their impact on daily life not only for patients but also for those around them. This situation upsets everything from health status to quality of life, friendships to family life, hobbies to professional life. What they have in common is that they systematically affect the social, psycho-logical and economic dimensions of the patient's life*" (WHO 2005, p. 15). The constant increase in NCDs cause pain and a large number of years of life with disability (GBD 2017 Disease and Injury Incidence and Prevalence Collaborators 2018) (Table 2.1).

People with NCDs represent around a third of the general population; for example, 22 million of the 66 million people in France. Their conditions are most often the result of environmental causes with a prevalence that increases with social, educational, or economic difficulties. NCDs largely result from harmful health behaviors.

The signs of entry into an NCD are less obvious than the case of an acute disease; they are insidious and their evaluation is less predictable (Frey and Suki 2008).

Patients, therefore, require care that is coordinated over time and that considers their needs, values, and preferences. They need to learn to self-manage their disease and to prevent complications. *"The therapeutic challenge is as much about learning to live better and in better health with the disease than gaining days of life at all costs. This work must be done in close collaboration"* (WHO 2003, p. 19). However, some medical doctors may consider that a chronic disease is stable when a treatment manages to balance a biological parameter in its targeted zone. Unfortunately, from the perspective of the patient and entourage who cope with the disease on a daily basis, this is not the case. Chronic pain, chronic stress, temporary discomfort, confusion, worrying signs, unexplainable manifestations, or allusive comments from a healthcare professional all make a patient's life very unstable. Many experienced symptoms and perceived reported outcomes remain variable over time (Ninot et al. 2010). The results showed that physical self-worth was less stable in moderate COPD patients than in same-age healthy adults. The higher fluctuations in the COPD

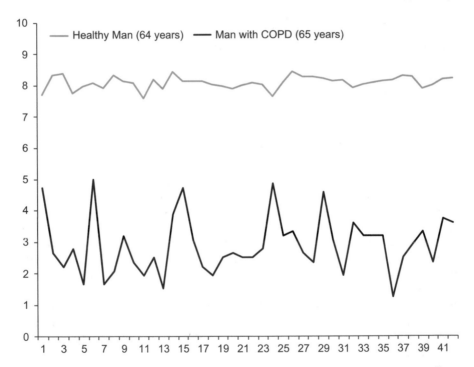

Fig. 2.1 Changes over a three-week period in perceived physical self-worth in two men, healthy or with a chronic obstructive pulmonary disease (COPD). Each participant completed an inventory on a single page of a personal notepad twice a day between 7:00 and 9:00 (AM and PM). The participants assessed their immediate, contextually based feelings of self-perception. They responded to the physical self-inventory using a visual analog scale (10-cm horizontal line) ranging from "not at all" (0) to "absolutely" (10). Reprinted from European Review of Applied Psychology, 60(1), Ninot, G., Delignieres, D., Varray, A., Stability of physical self: examining the role of chronic obstructive pulmonary disease, Figure 1, page 37, 2010, with permission from Elsevier

patients indicated that they resist less effectively to several endogenous and exogenous constraints. In other words, the patients seemed to be more sensitive to their
health status and environmental change. Many patients with COPD find it difficult
to cope with the distressing symptom of breathlessness. They are subjected to other
unpredictable, frightening, and disabling symptoms that make it hard to feel in control of their lives. Many patients develop a lack of confidence in their ability to avoid
breathing difficulties while participating in certain activities, however minimal the
physical demands of the activity may be. Greenier et al. (1999) stated that instability
in these dimensions has "psychological" meaning: it both promotes and is a product
of the fragility of feelings of self-worth. This statement could contribute to explaining why anxiety, health status, and depression increase the risk for hospital readmis-
sion in COPD patients (Maurer et al. 2008). This instability translates into a real
vulnerability that requires frequent listening and adjustments made by the patient,
their entourage and healthcare professionals; it is a key component of twenty-first-
century personalized medicine (Fig. 2.1).

**The Screen Screens the Care Relationship Between the Doctor and the
Patient**

Everyone may have experienced a tragedy in their life because of a medicine
that is perceived as hard line. The agony of a loved one can generate feelings
of relentless therapy, poor pain management, and false hopes. With rare
exceptions, these are not due to medical errors, but to a lack of explanation,
misunderstandings, or overly narrow regulations. These forms of suffering
testify to a medical system that judges its social utility in terms of a number
of years of life gained. The insistence on increasing lifespans has justified and
accelerated the technologization of medicine. Machines are essential for
establishing a diagnosis; screens are ubiquitous. The screen screens the care
relationship, and the doctor seems to interact more with his computer than
with his patient. He no longer watches. He does not listen anymore. He cannot
hear anymore. He no longer examines. He no longer feels. He has no time left.
He seems trapped in a system that asks him to produce ever more prescriptions and technical acts. He hopes the day when the algorithms, robots, and
drones will be better able to identify a disease than him, and to calculate the
best probability of the success of a treatment and to administer it. What will
remain of the doctor?

4.1   Slow and Natural Care vs. Fast and Artificial Biotherapy

More and more voices are being raised against biotechnological medicine, which no
longer takes time to listen to the true aspirations of patients and to understand their
way of life. They are against an impersonal and deaf medicine, a medicine adorned
with certainties about the future, a medicine remunerated by the number of treat-

ments delivered, a medicine that seeks the extension of the lifespan at all costs, and at all prices. More and more people are asking for a more natural, slower, gentler, more comprehensive, more holistic health approach – in summary, something more human.

4.2   Positive Health, Wellness Lifestyle

Faced with a world made violent by a fraction of humans, more and more voices are rising to demand a life in peace. Instead of resolutions on the first of the year that are quickly forgotten a few days later, more and more people are sharing – notably on social networks – about their life changes; for example, changes that bring them to live in the countryside, in a place that is detached from violence, where they can reconcile with themselves. The awareness of a finite world and a short human life makes the pleasures of simple, natural things more salient and is often linked to cultural, historical, and environmental roots. This approach focuses on the intimate, on singularity, and on getting rid of the ready-to-wear and ready-to-think that advertisers burn into our brains. This movement of positive health, of wellness, has only intensified with the tragic COVID-19 pandemic, which has forced more than half of the humans on the planet to live in prolonged confinement. This episode made us understand how much health depends not only on doctors, but also on our own behaviors, and to see the extent to which our choices and our behaviors impact not only our own health, but also that of our near and less close relatives.

4.3   End of Life

Debates over euthanasia ignite and divide opinion, both in the ranks of health professionals and among citizens and justice professionals. These debates are revived when sensitive cases occur, such as that of Vincent Lambert – a psychiatric nurse, quadriplegic and victim of irreversible brain damage following a road accident; or of the novelist Anne Bert – who suffered from amyotrophic lateral sclerosis (Charcot disease) and had to expatriate from France to Belgium in order to shorten her suffering and die with dignity. Patients often gather in committees or in federations that claim the right to be treated with the utmost dignity and to avoid suffering as much as possible. In some cases, parents, relatives, and families are able to say at what point decisions are made without consultation or consideration.

4.4   Health-Related Quality of Life

Personalized medicine requires knowing the patient's unique psychosocial situation, taking their personal preferences and health beliefs into account, and considering their values and goals as a means of improving a primary outcome – health-related quality of life. This concept is paradoxically recent, coming after the general definition of the quality of life that was disseminated by the WHO (1994).

Defining Quality of Life
In its broadest sense, the WHO defines the quality of life as *"an individual's perception of their position in life in the context of the culture and value systems in which they live and in relation to their goals, expectations, standards and concerns"* (WHOQOL Group 1994).

The WHO definition encompasses the individual's somatic state, physical autonomy, psychological state, psychological functioning, well-being, social interactions, professional activities, and economic resources. A more operational definition that is restricted to health corresponds to an *"aggregate of representations based on health status, physiological state, well-being and life satisfaction"* (MacKeigan and Pathak 1992). It excludes professional and spiritual areas, and it is a concept that is both subjective and multifactorial (Leplège, 1999). This definition approaches the concept of health that was defined by the WHO in 1947, giving it all its perceptual, contextual, intimate, subjective and singular value, and not just absence of disease.

Evaluating Health-Related Quality of Life (HRQL)
Health-related quality of life reflects a person's physical, psychological, and relational aspects. It results from a set of perceptions of health, personal well-being, and life satisfaction (Curtis et al. 1997). It is assessed, in particular, by scientifically validated questionnaires that are sensitive to change. These are used in clinical practice and in research as a marker for the success of a preventive or therapeutic action.
    Assessing health-related quality of life in any preventive or therapeutic use of NPIs:

- Considers the patient's opinion and feelings
- Testifies a shared decision
- Respects the patient as a person
- Quantifies change
- Does not seek to increase lifespan at all costs

Health-related quality of life (HRQL) provides an overview of how the patient feels (Ware 2003). Only one person lives daily with their disease; thus, what he or she has to say should guide health behaviors. Poor quality of life reflects pain, the loss of autonomy, a lack of knowledge about the disease, the behavioral mismanagement of the disease in routine or emergency situations, loneliness, a lack of support from friends and loved ones, and difficulty accepting the disease.

Today, various national and international strategies are calling for the consideration of the HRQL of patients with chronic diseases. For example, the French interministerial strategy 2007–2011 for "improving the quality of life of people with chronic diseases" was the trigger for concrete actions in favor of HRQL for patients with NCD (French inter-ministerial strategy 2007).

General practitioners are very sensitive to HRQL, even if they are not yet used to assessing it or of keeping track of it on their computers. Learning to do so would allow them to better understand the barriers and determinants of the quality of life as related to their patients' health. This would make them better caregivers because "the general practitioner is there to put people at the heart of medicine, to bring the patient relief, comfort, and hope from the first to the last day" (Queneau and de Bourguignon 2017).

The use of quality-of-life assessment as a supplement to objective clinical indicators is becoming increasingly vital in light of new and compounding questions about the effectiveness and appropriateness of many existing treatments. It also represents a paradigm shift in our approach to the operationalization and measurement of health outcomes. The measurement of the health outcomes of clinical interventions has become a cornerstone of health services research: it is essential for the assessment of the effectiveness and the appropriateness of healthcare interventions. Quality-of-life assessment is also increasingly popular among pharmaceutical companies, with most reporting that they have used some type of quality-of-life instrument in their clinical trials of drugs. Information about broader patient outcomes is also necessary in order to guide and empower patients so that they can make appropriate healthcare decisions (Bowling 2001).

5   Combining Solutions to Treat Complex Health Problems

It is not uncommon to find that a person suffering from a chronic disease has other health problems (WHO 2006). This includes symptoms such as recurrent chronic pain, fatigue, or stress episodes, as well as other syndromes like hypertension, heart disease, vascular disease, diabetes, cancer, osteoporosis, depression, anxiety, or neurodegenerative disease. Furthermore, despite what some organ specialists think, chronic disease does not stop at its symptomatic manifestations or at an isolated pathophysiological mechanism. From 50 years of age, 50% of people with NCDs also suffer from at least one other disease. After 65, this proportion rises to 80%. Also after 65, 62% of Americans suffer from at least two chronic diseases (Vogeli et al. 2007). This reality was further demonstrated by a study of a representative population which found that 42% of people with one chronic disease suffer from another health condition (Barnett et al. 2012) and that 23% of them have to face more than two additional pathologies. These patients have to cope with multiple condemnations.

Illustration

A person with one chronic disease acts as a breeding ground for the multiplication of comorbidities. This is due to:

- Residual sequelae of the initial treatments for the primary chronic disease (side effects of a surgery, for example).
- Complications of radiotherapy (skin reaction, for example).
- Side effects of a medication (nausea, for example).
- Nonadherence with medical prescriptions.
- Points of weakness that the body managed to mute before the disease (lower back pain that wakes up, for example).
- Sedentary lifestyle (increased sitting or lying time, for example).
- Lack of psychological resources (strategies to manage a stressful situation, for example).
- Nutritional deficiency (of vitamins, for example).
- Risk drivers (increased smoking, for example).
- Relational difficulties (family conflict, weak support from loved ones, prolonged work loss, loss of confidence in caregivers).
- Lack of social support.
- Lack of economic resources.

Within the context of NCDs a small problem can have serious consequences. From the biological to the environmental scales, the processes involved in these pathologies are complex, intricate, and the purveyors of systemic effects. On the biological level, researchers have observed numerous metabolic, cardiovascular, inflammatory, bone, muscular, and cerebral disturbances in patients suffering from a chronic disease (Fabbri et al. 2008). Psychologically, anxious and depressive phenomena are found to weaken patients (De Ridder et al. 2008). Cognitive disorders (attention, memory, analysis) also come into play. False beliefs persist and damaging consequences emerge on the social level: job loss, marital separation, family misunderstanding, insufficient financial reserves to cover the costs of increasingly expensive treatments and care, and the dispersal of the patient's network of friends. On the environmental scale, pollution, allergens, pesticides, and other toxic agents also more easily affect these people.

The development or presence of an additional disease can be even more serious and disabling than the initial disease. For example, 66% of patients with a fatal chronic disease, like chronic obstructive pulmonary disease (COPD), die from another disease (Domingo-Salvany et al. 2002).

These diseases combine exponentially rather than just compounding one another. In doing so they degrade the patient's overall state of health, causing more frailty and handicaps, and deteriorating their quality of life, which then causes increasing collateral damage within their families. Medical consultations, examinations, and treatments increase in frequency and complexity as care becomes more complicated. Hospital stays become more lengthy and medical opinions less unanimous. The future is more uncertain.

These determinants combine, overlap, influence, and build on each other, all the while weakening the patient ever more in their daily life. For professionals, as the situation becomes more complex, it becomes more challenging to offer an optimal care approach, requiring more time for reflection and interdisciplinary exchanges. The evolution of the patient's state of health becomes less predictable (Frey and Suki 2008) and their prognosis more uncertain.

As if that were not enough, two factors that play a role in accelerating the risk of comorbidity further darken the picture: age, on the one hand; and social insecurity, on the other. The effects of aging are always at work, inviting the development of other health problems. Additionally, people who are without sufficient financial resources, isolated, and/or are of a low sociocultural status are more vulnerable to comorbidities. In cancer, for example, they benefit from less care and support offerings (Corroller-Soriano et al. 2008). They have also been found to treat themselves less, to treat themselves less well, to consult medical practitioners less, and to have a harder time responding to administrative requirements.

> **Systemic Consequences Increase the Cost and Complexity of Care Strategies**
> A patient with one chronic disease consults the doctor four times a year on average. In the case of five or more diseases, the number of consultations increases to 14 times a year (Vogeli et al. 2007). Comorbidities imply more serious consequences on health, compounding the complexity of implementing, adjusting, and coordinating treatments. The cost of care is skyrocketing (Vogeli et al. 2007), both for the healthcare system and for patients.

Care strategies for patients with comorbidities challenge the organization of a health care that has been almost exclusively focused, since the last century, on treating one disease at a time – usually in an emergency situation (e.g., infection, exacerbation, accident, acute crisis). The increase of the complexities, uncertainties, and singularities of each case requires comprehensive and adaptive programs that combine care, prevention, education, and social support that is adapted to individual preference and lifestyle. These are the programs that are deployed for most chronic diseases today. Pivotal clinical trials for each chronic disease have justified the need for both their implementation and systematization for medical, social, and economical reasons. They are, at the same time: (1) beneficial for the patient's health, autonomy, longevity, and quality of life; (2) less at risk of causing side effects; and (3) more likely to generate savings by limiting the expenses of nonprogrammed care. In order to have greater medico-economic impacts, these care strategies are increasingly proposed in the early stages of the disease and no longer only at an advanced stage.

A Pivotal Trial for COPD Comprehensive and Integrative Care Published in 2000

Chronic obstructive pulmonary disease (COPD) is ranked as the fourth-leading cause of death worldwide. Many people confuse COPD symptoms with those of bronchitis, or attribute them to fatigue or aging. COPD is generally caused by active tobacco smoking or exposure to secondhand smoke. It is an incurable disease. Dyspnea, its main symptom, is akin to an anxious feeling of breathlessness. A British randomized controlled trial published in the journal *The Lancet* tested the efficacy of a multidisciplinary program consisting of three weekly hospitals sessions over the course of a month and a half (Griffiths et al. 2000). These sessions were personalized, lasting two hours each; they included endurance exercises (indoor walking and biking), muscle training with resistance elastics, smoking cessation for smokers, and therapeutic education (disease and stress management). Participants were encouraged to pursue these activities at home. All participants were randomly assigned to either the rehabilitation group or the control group; the control group followed traditional recommendations and treatments. The trial assessed the main clinical, medical, and psychosocial characteristics of the participants, as well as the number of their hospital days and home consultations. At the end of the program, the results of the trial showed that the comprehensive and integrative rehabilitation group benefited more positively than the control group. This positive impact lasted throughout the year following the study, with reduced dyspnea, improved exercise tolerance, reduced depression symptoms, and improved HRQL. Furthermore, the results evidenced a significant difference in the number of hospital days in the year following the rehabilitation program (10.4 days on average vs. 21.0 days for the control group). A 6-week rehabilitation program offering three 2-hour sessions per week improves the clinical factors, functional autonomy, and quality of life of COPD patients; it also reduces the number of hospital days by 50%. The total number of sessions (18) represents a minimum threshold needed to obtain significant benefits. Using a multidisciplinary approach allows professionals to effectively deal with all facets of COPD. An incremental cost/utility analysis concluded that the program is cost-effective after one year and is likely to result in financial benefits to the health service (Griffith et al., 2001). The comprehensive and integrative rehabilitation program improves the efficiency of overall healthcare use with respect to reducing days spent in hospital and the number of visits to the emergency department.

A comprehensive and integrative strategy that includes NPIs will obtain better biological and psychosocial results through combined treatments. Going further, this strategy will be more efficient due to its inclusion of a multidisciplinary team, which allows it to maintain an extensive, holistic regard. On the patient's side, this interactionist and situationist strategy allows them to no longer be passive ("patient") but rather to become the main actors in their health (e.g., embodiment, empowerment, disease management); in other words, to assume a part of the responsibility in their treatment results.

6   The End of the Myth of the Rational Human

In health prevention, it was long thought that it was enough to merely inform or to indicate what good behavior is for it to be applied. Even though 99% of the population knows that physical inactivity, smoking, alcohol, and unhealthy eating affect health, 33% of them are unable or unwilling to change their behavior. Half of those who do manage to change their behavior give up after six months. In short, knowledge does not sustainably modify health behavior. For example, a February 2007 survey conducted by the French Institute for Prevention and Health Education on the impact of health messages about food showed that 71% of the people questioned had memorized the health messages a few months after their implementation; nevertheless, 64% of the respondents said that this knowledge has not changed their habits. Only 9% stated that they had started to modify their diet, and only 4% had changed their buying habits concerning the brands and products that were indicated. Humans are not rational about their health.

> **The Myth of Human Rationality and Decision Making**
> Richard Thaler, the 2017 Nobel Prize winner in economics, definitively dashed all hopes of seeing humans as rational beings. He showed how irrationality and social preferences affect individual decisions and behaviors. He also showed that the human aversion to loss explains why people value what they already own more than what they don't. Applied to health prevention, the goal of preventive strategies should not "only" be to inform and to raise awareness about virtuous behaviors, rather to also encourage people to adopt them. Empirical, experimental, and neuroscientific data have all showed that psychological and social biases characterize human decision making, thus refuting the rationality hypothesis of human beings – it is a myth.

A major pitfall of all nonspecific primary health prevention campaigns is that they generally reach populations that are already aware of good health practices – those who need them the least – while they miss their target of people who are at risk or who may be in denial of their health problems. Another pitfall of this type of health prevention campaign is that they may be overly punitive. A questionnaire that was sent to smoking cessation experts from the European Network for Smoking Prevention (ENSP) in 28 European countries estimated the contribution of all tobacco control measures, based on 100, as follows:

- Increases in the price of cigarettes: 30.
- Smoking bans in public places: 22.
- Awareness campaigns: 15.
- Bans on all tobacco advertising and promotion: 13.
- Health warnings: 10.
- Strategies to help addicted smokers quit: 10.

Using NPIs in health prevention becomes obvious, since they act as much on our rationality as on our irrationality and emotions. Thus, they promote both target behavior changes and long-term maintenance.

With disease therapy, human irrationality is worse. When our lives or the lives of our loved ones are in danger, buried anxieties and magical thoughts arise. Causal links are found in order to explain a state that goes beyond all logic, inciting superstitions or rituals that confuse doctors. For example, once before a risky crossing, the families of sailors would hang a marble plate where their prayer was inscribed on the wall of a church. Parents shaken by their child's illness did the same. The Saint-Roch Church in Montpellier preserves multiple traces of these men and women who wanted to put "chance" on their side and to believe in a positive outcome.

7   The Digital and P4 Medicine Revolutions

We now believe that using high technology to produce and integrate enormous data sets leads to an improved understanding of human health.

7.1   The Quantified Self

Many of the findings of studies evaluating NPIs such as psychotherapy, adapted physical activity programs, or diets have been challenged due to their exclusive use of questionnaires. Some authors have underscored the methodological limits of these self-questionnaires and their sensitivity to different biases (e.g., memory bias, social desirability bias, contextual bias, contagion bias, influence of the evaluator, environmental distraction).

Today, however, the answers to these questionnaires can be cross-referenced with behavioral data (e.g., number of daily steps), biological data (e.g., heart rate, blood sugar level), and economic data (e.g., care expenditures). This compiled, cross-checked, and corrected data gives more credibility to these analyses.

7.2   Data Management

In the space of 20 years, the cost of a randomized controlled trial has been divided by 10 (administrative formalities, digital records, data collection, data transmission, data verification, analyses).

7.3   Big Data Analysis

Twenty years ago, access to medical research data was restricted and rarely exhaustive. Information on professional practices was kept in a restricted circle. Storage space was limited. Today, search engines and databases have made health data more easily accessible and available in multiple formats (text, image, video). The possibilities for analysis, cross-checking, and historical feedback are now endless.

7.4   P4 Medicine

A new healthcare system called P4 medicine – predictive, preventive, personalized, and participatory – emerged in the beginning of this century. New technologies and algorithms contribute to the implementation of this medicine, and consequently to the advent of NPIs.

> **P4 Medicine: Predictive, Preventive, Personalized, and Participatory**
> *"Ten years ago, the proposition that healthcare is evolving from reactive disease care to care that is predictive, preventive, personalized and participatory was regarded as highly speculative. Today, the core elements of that vision are widely accepted and have been articulated in a series of recent reports by the US Institute of Medicine. Systems approaches to biology and medicine are now beginning to provide patients, consumers and physicians with personalized information about each individual's unique health experience of both health and disease at the molecular, cellular and organ levels. This information will make disease care radically more cost effective by personalizing care to each person's unique biology and by treating the causes rather than the symptoms of disease. It will also provide the basis for concrete action by consumers to improve their health as they observe the impact of lifestyle decisions. Working together in digitally powered familial and affinity networks, consumers will be able to reduce the incidence of the complex chronic diseases that currently account for 75% of disease-care costs in the USA"* (Flores et al. 2013).

8 Early Diagnosis

The earlier the diagnosis, the better the chances of recovery. A mammogram detects small abnormalities before the first sign of cancer. Early diagnosis allows better patient care and increases their chances of recovery. People with hypertension are often asymptomatic until they develop end organ damage. Proactive strategies must be adopted for the early detection of hypertension, with an emphasis on primary health care (WHO 2014). Hypertension screening and early detection improves the efficiency and effectiveness of the management of several modifiable factors that contribute to its high prevalence rates (e.g., eating food containing too much salt and fat, the inadequate intake of fruits and vegetables, overweight and obesity, the harmful use of alcohol, physical inactivity, psychological stress, socioeconomic determinants, and inadequate access to health care).

> **Emphasis on Early Detection**
> *"Evidence of benefit for lowering blood pressure levels at or above 160/100 mmHg with drug treatment and non-pharmacological measures is very clear"* (WHO 2014, p. 72); however, *"worldwide, detection, treatment and control of hypertension are inadequate, owing to weaknesses in health systems, particularly at the primary care level"* (WHO 2014, p. XIV).

8.1 Breast Cancer

In 2013, Angelina Jolie decided to have her breasts removed (mastectomy) because of a simple blood test that revealed that she was carrying a BRCA1 genetic mutation. Her decision to get tested was motivated by the fact that several women in her family had died of cancer, such as her maternal aunt. Two years later, in agreement with her doctors, she underwent another surgery to have her ovaries and fallopian tubes removed. She then wrote a brave article in *The New York Times* in order to educate women with a BRCA1 or BRCA2 genetic mutation about the risks of breast and ovarian cancer. These risks are significant before the age of 50; 72% for carriers of BRCA1 and 69% for carriers of BRCA2 (Kuchenbaecker et al. 2017).

8.2 Colorectal Cancer

Early diagnosis is important for colorectal cancer. When detected and treated at an early stage (stage 1, superficial cancer of the intestinal wall), healing is attained in 90% of cases and the 5-year survival rate is 94%. When detected late (stage 4, spread of metastases to other organs), the chances of recovery decrease considerably and the 5-year survival rate is only 5%.

8.3 Skin Cancer

Skin cancer is another example. Detected in time, it heals; treatment involves removing the lesion under local anesthesia. Detected late, when the melanoma is thick or metastasized, the chances of recovery are low due to the current ineffectiveness of treatments beyond removal.

8.4 Alzheimer's Disease

The characteristic signs of Alzheimer's disease in the hippocampus do not appear suddenly. They result from a slow process of degradation, in which the first signals can go unnoticed for more than 20 years in some cases. The patient tends to blame memory difficulties on advancing age, stress, inattention, or transient fatigue. As a result, relatives are often the first to recognize its symptoms. Neuropsychological tests allow doctors, neurologists, and geriatricians to confirm the diagnosis and to estimate the severity with the help of a specialized neuropsychologist. In its early stages, NPIs are useful for delaying the progression of the disease, for developing compensatory strategies, and for minimizing associated handicap situations (Prince et al. 2011).

In the future, many diseases will benefit from early diagnosis, advanced medical imagery, neuropsychological tests, behavioral analyses, and lifestyle questionnaires. Cancers, heart disease, neurodegenerative diseases, multiple sclerosis, osteoarthritis, COPD, and type 2 diabetes (including prediabetes) are some convincing examples.

9 The Collateral Damages of Biomedical Treatments

What side effects from some biochemical treatments! What complications and disabilities left by some surgeries! How many hospitalizations that could have been avoided! How many biotherapy scandals that have caused hundreds of deaths and disabled thousands of people! How many clinical trials rigged by administrative, recruitment, or statistical tricks! How many unpublished negative studies! How many diseases invented solely to increase the sales of a molecule! How many court cases revealing the actions of lobbyists! What medical training that continues under an influence leading to over-prescription! What conflicts of interest in authorizations giving market access to unsafe treatments! What resistance from the authorities who refuse to hear complaints from patients taking drugs like Levothyrox!

9.1 Questioning the Model of "a Single Curative Drug for Every Disease"

Avastin, for cancer, has caused blood clots, heart failure, and intestinal perforations. Avandia, for diabetes, has caused heart attacks. Torcetrapib, for heart disease, has caused death. An increasing number of people with chronic diseases around the world are testifying to the failure of drugs to cure complex pathologies. This makes people more aware of side effects and inappropriate prescriptions. Peter Gøtzsche is the directing head and 1983 co-founder of the Scandinavian Center of the Cochrane Collaboration; he worked for the pharmaceutical industry from 1975 to 1983. From 1984 to 1995 he worked for the Copenhagen hospital and was appointed professor at the University of Copenhagen in 2010. His book, published in 2015, explains that about 200,000 patients die each year in the United States due to the side effects of drugs. In half of those cases, the drugs are prescribed correctly. In the other half, deaths are the result of overdoses on the part of either doctors or patients. His book also warns about the over-treatment of healthy people.

Court cases leading to the conviction of pharmaceutical companies publicize the collateral damage of pharmaceuticals. They encourage authorities to investigate more thoroughly and doctors to prescribe and monitor pharmaceutical use in a more careful manner. In parallel, the WHO is concerned about the misuse of antibiotics causing antimicrobial resistance in populations that are difficult to treat afterward (WHO 2015). Controversy also surrounds the use of very expensive cancer treatments that only extend life for a few months and have no real impact on the patient's quality of life. No drug has yet proven efficient or effective for Alzheimer's disease. Today, there are discussions of no longer offering cholesterol-lowering statins to a large population, even though their use seemed encouraging only a few years ago.

Intoxication with Pain Medication

Opioids are substances with known analgesic effects. They constitute a large pharmacological class that brings together the natural alkaloid molecules derived from the *Papaver somniferum* plant (morphine and codeine), semi-synthetic molecules (created by the chemical modification of an opiate such as heroin or oxycodone), and synthetic molecules (e.g., meperidine, methadone, fentanyl). Their use, either in medical routines or illicitly, is very frequent. They can cause severe states of intoxication, which can lead to serious consequences and death. They all have a simultaneous activity on several receptors and several classes of receptors, which makes the prediction of their expected global effects more complex. Their impact on breathing is the main clinical sign, since they decrease both the frequency and range of breathing, reduce the reactivity to hypercapnia and hypoxia, interfere with the cough reflex, and, in some cases, cause muscle stiffness in the chest and abdominals. Additionally, they can decrease muscle tone in the upper respiratory tract, contributing to lung obstruction. How about using NPIs to relieve pain?

9.2   Controversies About Surgeries and Medical Devices

Considered as a quick and radical solution, obesity surgery – called bariatric surgery – is no longer unanimously accepted. Bariatric or obesity surgeries are now being debated and their real effect on extending lifespan is disputed. Following these surgeries, deleterious effects on the daily lives of patients appear in the medium term. In bariatric surgery, for example, 60,000 patients are operated on in France each year. Compared to the population, this rate is higher than that of the United States. Bariatric surgery modifies the digestive system by reducing the volume of the stomach to limit the amount of food consumed (e.g., by placing a gastric band or amputating part of the stomach) or by shortening part of the stomach and small intestine so that ingested food is absorbed less during its passage through the digestive tract (e.g., a bypass). Late-term complications of surgical or nutritional origin that are more or less serious can occur, such as neurological damage, excess skin, muscle breakdown, nutritional and vitamin deficiencies, and identity troubles. Patients are rarely informed of all of the potential consequences of these surgeries and must also pay excessive fees.

A variety of other surgical practices are also currently being discussed, especially for cases that are not serious. For example, the stenting of healthy people without high risk of heart disease and pain that follows a routine medical consultation leading to a series of in-depth examinations. Once in place, they must take medication for life to prevent the real risk of clotting in the stent. Thus, the people receiving this medical device become chronically ill patients under continuous surveillance. Other medical devices can augment medical and economic costs because of patient misuse, even if their efficacy is proven under experimental conditions. For example, this is the case for ventilation machines that are used for patients suffering from obstructive sleep apnea syndrome. This syndrome is characterized by the repeated interruption of breathing due to obstruction of the upper airway during sleep. Its prevalence in the general adult population has been found to range from 6% to 17% and to be as high as 49% at an advanced age. These patients have an increased risk of cardiovascular illness, car accidents, and use of healthcare resources. The gold standard for treatment is a medical device called a continuous positive airway pressure system. The machine improves daytime sleepiness and cognitive function, and reduces mortality, sympathetic neural activation, and blood pressure. Continuous positive airway pressure therapy uses mild air pressure to keep the airways open, ensuring that the airway doesn't collapse and impede breathing during sleep. The machine needs to be used every night, and it includes a mask that covers the nose and mouth, a tube connecting the mask to the continuous positive airway pressure machine's motor, and a motor that blows air into the tube. Patient health and survival benefits depend on adherence to use of the device (defined as use for at least four hours per night, for at least 70% of nights, and for at least 30 days in the first three months after prescription). Some studies show that as much as 83% of patients are nonadherent to its use, making poor adherence a major problem with this treatment in the short term that increasingly worsens in the long term.

9.3   Mistrust of the Medical System

It is not a question of demonizing an industry that cures many health problems and has allowed considerable therapeutic progress in the last century. It is not about exposing nonexistent conspiracies. Simply, it is important to shed light on the increasingly fierce competition between pharmaceutical companies that are under financial pressure in a market in which innovations are rare and increasingly expensive to achieve.

The problem is that the denunciation of certain practices and of certain firms brings the whole industry into disrepute in the eyes of public opinion. As a result, it is clear that more and more people are turning away from biochemical or biotechnological treatments as a first recourse, and this trend is no longer limited to society's most precarious or most radical citizens.

It is becoming increasingly difficult to still wonder why patients and professionals are increasingly turning toward natural remedies, genuine care recipes, and traditional medicines. Scandals, true or false, real or only exaggerated by journalists looking for sensationalism, are encouraging the development and use of NPIs as never before.

10   Intervention Studies in NPIs

The high cost and limited success in developing new pharmaceuticals in the twenty-first century has led to suggestions that a different, broader approach to dealing with disease is required. Such an approach could include delaying the onset of disease by controlling modifiable lifestyle factors, such as smoking, obesity, and exercise (primary prevention); by the timely identification of the disease (secondary prevention); and by implementing post-diagnostic interventions aimed at improving quality of life and delaying disease progression (tertiary prevention). Increasing importance is being given to, and a growing number of studies are investigating new diseases and novel strategies for treating, managing, and supporting people with chronic diseases and their caregivers.

Throughout the last century there was a paucity of high-quality evidence for NPIs. While observational studies reported NPI benefits on different indicators – health status, quality of life, costs – unfortunately, these observations did not qualify as sufficient scientific and medical evidence to undoubtedly recommend their systematic implementation (Boutron et al. 2008; Ernst 2011; Ninot 2013). To obtain these forms of evidence, it has been necessary to develop and use evaluation protocols that are rigorously framed and followed. However, some authors and academic organizations remained skeptical to the point of contending that it is impossible to evaluate these practices with rigorous protocols (see Chap. 8).

In the last 20 years, major improvements have been made in the standardization and quality of clinical trials for NPIs. As a result, the number of intervention studies investigating the overall effectiveness and the cost-effectiveness of NPIs has been rapidly growing. These interventions target a wide spectrum of diseases and symptoms, going beyond biological deterioration, as is the limitation for most studies of biochemical drugs or surgeries, to include psychological and behavioral aspects, functional abilities and activities of daily life, social participation, and cost-savings, among others. Factors unrelated to disease, such as low levels of education, the presence of psychological distress, and low perceived social support – all of which are amenable to NPIs – have been shown to contribute to the risk of developing diseases and, therefore, to be potential targets for therapeutic interventions. Moreover, in developing countries, where financial and logistical resources constrain healthcare delivery, targeting these risk factors with low-cost, patient-driven, and peer-led NPIs is an attractive and sustainable option.

Research and innovation on NPIs are experiencing unprecedented growth worldwide. Public and private research funding is becoming more substantial, diversified, and ambitious. Each year, there are over 100,000 publications on the subject in peer-reviewed scientific and medical journals.

Experimental studies evaluating the overall effectiveness, risks, and cost-effectiveness of NPIs are called "intervention studies" (also called comparative studies, controlled studies, before-after studies, trials, clinical trials). The simplest case, the before-after study, verifies if an NPI causes statistically significant changes in the average related to one or several health outcomes in a group of participants before and after its administration (Fig. 2.2). A controlled study with the same design compares a group of volunteers testing an innovative NPI (called the experimental group) with a group of people following a conventional care protocol or an intervention with no effect (control group). Researchers verify the hypothesis that the NPI will have a statistically higher benefit on biological, psychological, social, and/or economic aspects related to health markers than the standard or control intervention at the end of the study (Fig. 2.2). A randomized controlled trial (RCT) compares an "intervention" group, testing a new NPI, with a "control" group following routine care recommendations or taking a placebo. Participants are randomly assigned to one of the two groups. The choice of group assignment can be made either with the knowledge of the researchers and clinicians (single-blind) or without the knowledge of researchers and clinicians (double-blind). By controlling all variables and ensuring the homogeneity of the groups at the beginning of the study, it is possible to isolate the role that the new NPI may play when considering the differences obtained on measured indicators at the end of the study. Thus, this type of protocol formally establishes a direct causal relationship between the NPI and measured improvements (Fig. 2.2). The RCT study model provides the highest level of evidence for the effectiveness of a new therapy or new health prevention intervention according to the principles of evidence-based medicine (see Chap. 8).

The results of these studies are published in scientific and medical journals with independent peer-review committees. Algorithms and intervention descriptions can be used to better identify relevant articles and best matches for queries.

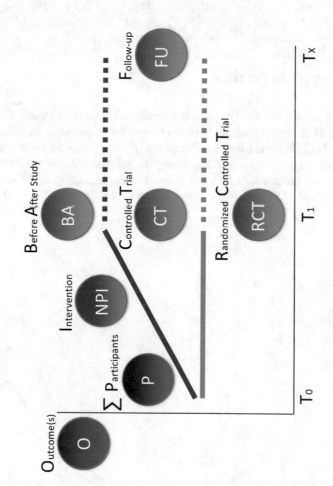

Fig. 2.2 Design of before-after, controlled, and randomized controlled studies

Humanity's First Clinical Trial Evaluating an NPI
On an English military boat cruising off the Bay of Biscay in 1747, Dr. James Lind decided to experimentally assess a remedy for the scurvy that was ravaging the sailors of the time. He divided 12 patients with scurvy into six groups of two. He gives each group a different substance and ensured that their nutrition was identical. The substances tested were cider, sulfuric acid, vinegar, a concoction of herbs and spices, sea water, and oranges and lemons. Only the last group recovered from scurvy. In 1757 he published the "Scurvy Treaty" that would save thousands of sailors around the world.

10.1   A Request on PubMed

The most well-known and freely accessible health database is PubMed, which was developed and is maintained by the National Center for Biotechnology Information (NCBI) at the U.S. National Library of Medicine (NLM). It provides relevant trends and proportions for studies about NPIs. Using the keyword "NPIs" on PubMed on June 7, 2020, we obtained 22,883 article references, including 2087 results for "clinical trials" and 1651 results for "randomized controlled trials" (Fig. 2.3). The

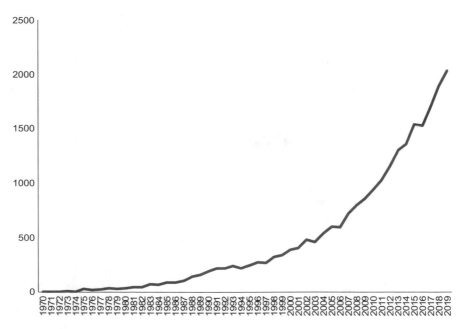

Fig. 2.3 Cumulated publications identified in PubMed with the term "non-pharmacological OR nonpharmacological OR non-pharmacological OR nonpharmacological OR non-drug OR nondrug OR non-pharmaceutical OR nonpharmaceutical" (query: June 7, 2020)

number of article publications has increased six-fold over the past 20 years. This exponential acceleration demonstrates the growing interest in NPIs in the field.

Another sign of interest in NPIs is the astonishing proportions that resulted from this PubMed query: the number of reviews (8944) is four times more than that for clinical trials (2087). This proportion reveals that there is more concern for delivering opinions than for producing experimental data. Once again, things are changing, since in the last decade there was a higher growth in clinical trials than reviews.

Several reasons explain why this PubMed request does not reflect the real number of scientific publications assessing the health benefits and risks of NPIs worldwide. The term "NPI" and its equivalents are not universally used in the titles, keywords, or abstracts of NPI articles, and terms are the main source for the algorithms. In many cases authors use either the specific name of the NPI being evaluated ("MBSR" for example), its subcategory ("psychotherapy" in this example), or its general category ("psychological intervention" in this example). Moreover, PubMed focuses primarily on biological publications, with less focus on psychological, educational, social, and economic studies in its database (the PubMed website states, "citations in PubMed primarily stem from the biomedicine and health fields, and related disciplines such as life sciences, behavioral sciences, chemical sciences, and bioengineering"). Last, but not least, English is the predominant language of PubMed. No database fully covers all available NPI studies at this time.

10.2   A Review of the Reviews Published in 2019

Going further in our exploration of PubMed, we wanted to know the number of narrative reviews, systematic reviews, and meta-analyses published in the international literature on NPIs in 2019, as well as their thematic coverage in the health field. As such, on March 1, 2020 we conducted a systematic review identifying all review articles published in 2019 in a peer-reviewed scientific or medical journal. We included all narrative reviews, systematic reviews, and meta-analyses looking at the benefits and risks of NPIs on a health determinant. The review could focus on prevention or care, whether it was or was not in addition to biomedical treatments. We searched articles in well-known health databases including PubMed, Cochrane Library (Central), ScienceDirect, and Web of Science. We also searched on Google, Google Scholar, ResearchGate, and Academia, eliminating redundancies if necessary.

We excluded articles that did not cover the thematic totality of NPIs; in other words, articles focusing only on a single theme (e.g., psychology, education, exercise, mind-body practice, nutrition, digital medicine, herbal medicine) or on a single medicine (e.g., traditional Chinese medicine, Ayurvedic medicine). We also excluded summaries without published articles, seminar reports, poster summaries presented at congresses, editorials for journals, books, book chapters, reports from academic societies (e.g., medical societies, health authorities), and student reports (e.g., internship report, PhD). We accounted for erratum by systematically considering the latest version.

The results indicated that there were 301 review articles published in 2019, all of which appear in the references in Blog en Santé (http://blogensante.fr/en/2020/03/02/301-revues-sur-les-inm-de-litterature-publiees-en-2019/): 83% on PubMed (250); 1% on Central (4); 7% on ScienceDirect (23); 3% on Google Scholar (8); 1% on ResearchGate (2); 1% on Academia (3); 2% on Google (6); and 2% on Web of Science (5). The results included 177 collective expertise articles, 52 systematic reviews, 43 meta-analyses, 15 individual expertise articles, 9 narrative reviews, 3 reviews of systematic reviews, and 2 reviews of meta-analyses. The majority were in English (281), followed German (6), Spanish (6), French (2), Russian (2), Polish (2), Chinese (1), and Hungarian (1). Two hundred eighty-two publications presented results and 19 presented protocols. The reference list is available on my blog (Ninot 2020).

The first authors of the articles came from 49 different countries, with a decreasing order of importance: the United States, England, Italy, Australia, and China (Fig. 2.4).

The articles relate to 51 health determinants, the first 20 of which are presented in Fig. 2.5. Pain treatment concerns 19% of the total review articles identified. Chronic diseases, addictions, problems related to aging, and prevention constitute a large part of the health topics addressed by these 301 literature reviews on NPIs.

The significant number of reviews covering the whole spectrum of NPIs being published in scientific journals around the world supports the idea that these evidence-based practices should be called "*non-pharmacological interventions*." The results of our enquiry suggest that a global approach going beyond a single category is the most relevant way to understand and implement NPIs for health. Several authors recommend using a combination of NPIs to gain greater health benefits in the short-, medium-, and long terms. Every person should have the right to put the odds on their side by using multiple treatment solutions, which requires a real change in medical habits. It also requires the creation of a transparent and rigorous decision-making process and implementation strategy for complements to biomedical treatments.

Last, but not least, it is worth underscoring that the PubMed database included 83% of the published NPI reviews identified. A person using this system must keep in mind that this database is not exhaustive, especially not for NPIs due to scientific reasons (its focus on the biological rather than the psychosocial), medical reasons (its focus on therapy rather than prevention), social reasons (its focus on the western world with journals mainly in English), and political reasons (its legitimate focus on American research and the protection of American industrial interests in line with it being entirely funded by U.S. government departments).

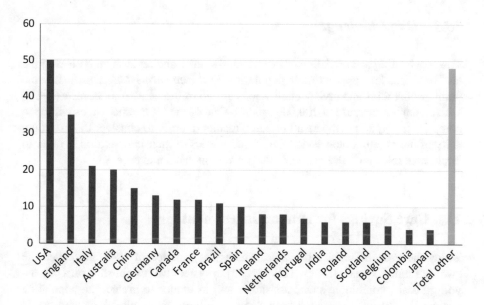

Fig. 2.4 Country of the first authors of the NPI reviews published in 2019

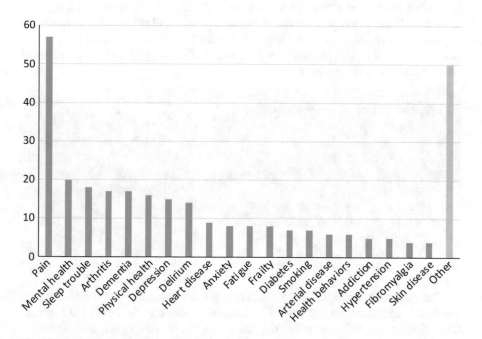

Fig. 2.5 Health determinants targeted by one or more NPIs

10.3   One Challenge

The accelerating accumulation of intervention studies and reviews on NPIs over the last two decades supports their popularity. The *Plateforme CEPS* estimates the number of studies and review publications at two million and 10,000, respectively, with an annual increase of 100,000 articles. One of the future challenges will be to compile, select, and compare all research results in order to consider the heterogeneity of their methodologies and the characteristics of their intervention in order to implement relevant NPI treatments for prevention, care, and cure.

11   Care Savings for Human Investment

Today the French Health *Assurance*, which was founded in 1945 and called the "welfare state," is under attack. This egalitarian system was established on a pay-as-you-go basis, combining social assistance and insurance to protect the population against the "risks" associated with old age, illness, unemployment, and social hardship.

Unfortunately, the slowdown in economic growth, the late entry of young generations into the workforce, the rise in unemployment, and the difficulties in financing social protection are accumulating, compounding with the new health and social needs (social exclusion, population aging, chronic disease) and with the exponential costs of treatment and hospitalization.

> **The End of Hospital-Centrism**
> Even though the French are attached to their health system, it is clear that its debts accumulate, despite benefitting from one of the highest percentages of GDP for health spending in the world (11%, 230 billion euros per year). It must be said that 50% of these expenses go to hospitalizations and only 2% go to prevention. Hospital services are saturated, in particular emergency departments that receive 20 million visits per year, 85% of which are cases that could have been managed by physicians. Getting France out of hospital-centrism is urgent.

The French national healthcare services are diversified, overcrowded, poorly regulated, and unevenly distributed to such an extent that, in 1996, the National Objective for Expenditure on Health Insurance was created by government order in order to control spending on care, including hospitalizations dispensed in private or public establishments and medico-social centers. A spending objective is set each year by the Social Security financing law and voted on by the Parliament. The French Health Insurance agency also publishes an annual report on its spending control proposals. The contribution of NPIs is encouraged in the areas of prevention (smoking cessation methods, for example) and day-to-day disease management (therapeutic education programs, for example).

In the face of increasingly constrained resources, determining which preventive and therapeutic treatments are most efficient is necessary: "Careful analysis of the costs and benefits of specific interventions, rather than broad generalizations, is critical" (Cohen et al. 2008, p. 663). For example, the authors of a study cited here found that cognitive-behavioral interventions are cost saving in patients with Alzheimer's disease. This type of analysis could identify not only cost-saving preventive measures but also preventive measures that deliver substantial health benefits relative to their net costs. Such analysis could also identify treatments that are cost saving or highly efficient (i.e., cost effective). These types of findings underline the importance of shifting practice toward a more cost-effective delivery of health care, a realistic way of achieving better health results. "*Conduct careful analysis to identify evidence-based opportunities for more efficient delivery of health care - whether prevention or treatment - and then restructure the system to create incentives that encourage the appropriate delivery of efficient interventions*" (Cohen et al. 2008, p. 663).

This path leading to a better reflection on the control of health expenditure and a better traceability of the care actions will allow the reallocation of investments into strategic areas such as innovation and the reorganization of care. These evolutions, which lean toward prevention and proximity care, will undoubtedly benefit the NPI sector.

12 An Ecosystem Full of Promise

Biomedical treatments are highly regulated and monitored by health agencies such as the European Medicines Agency (EMA) in Europe and the U.S. Food and Drug Administration (FDA) in the United States. NPIs represent a growing and attractive market requiring agile and local regulation.

12.1 Less Regulation, More Freedom, More Innovation

In the NPI ecosystem, users claim to be treated differently, emphasizing the benefits of the freedom of choice and medical pluralism that is delivered in a context of mixed behavior and culture. Until now – with the exception of medicinal plants that are quite similar to drugs – policies about NPIs and their professionals remain weak in Europe and are almost entirely absent in certain countries such as France (Wiesener et al. 2012). The European research program CAMbrella (2012) underlined that since the treaties of Rome and Lisbon left each state the freedom to organize their own health systems and health legislation, national governments are reluctant to open up the healthcare offerings and to harmonize care throughout the European territory. Furthermore, the 2014–2023 WHO plan (2013b) advocates for the integration of complementary medicine "into the conventional health system" via "appropriate national policies" (WHO 2013b, p. 44). However, practices vary widely from country to country with certain modalities regarded differently depending on the culture, understanding, and accessibility of conventional medicine.

In general, most NPIs are tolerated. This allows for the step-by-step building of an NPI ecosystem as a bottom-up process between two independent sectors: biomedicine and consumer products or services. The world's greatest innovations have typically occurred at the intersections of distant fields, industries, and cultures, and the NPI domain offers important opportunities for innovation as a means of delivering relevant healthcare offerings.

12.2   Solutions for the Growing World Wellness Economy

The use of NPIs is becoming widespread all over the world. They are the key elements of wellness. The WHO estimates the value of world wellness market at $78 billion. The Global Wellness Institute put this figure at $360 billion due to the inclusion of complementary medicines in 2017; when other well-being activities are taken into account, this figure further rises to $4.2 trillion (Global Wellness Institute 2018). The U.S. National Center for Complementary and Integrative Health (NCCIH) has invested $2.5 billion in research and information since 1998. Europe is starting to take an interest, and China, Germany, Japan, and Switzerland are all proactive players in the field (Chap. 7).

The NPI economy mirrors these shifting priorities, along with the growing recognition of the critical impact that behaviors and the environment have on our health and well-being. It is interesting to note that the NPI offering is not limited to the medical sector, but also extending to personal and community life, professional life, and touristic life. All of these sectors are essential components of the "NPI ecosystem" that nurtures a lifestyle characterized by well-being and longevity. The potential of and the opportunities offered by this ecosystem are of great interest to consumers, entrepreneurs, and investors. In the span of only a few years, there has been a rapid increase in acquisitions, partnerships, horizontal/cross-category expansions, and an emergence of new business models across all sectors of the wellness economy (Global Wellness Institute 2018).

The NPI ecosystem stands to profit from major investments in the related domains of the tourism, food, insurance, and technology industries, as well as support from academic organizations (research, development, innovation) in order to provide successful solutions that change our lives and provide sustainable employment for many in the next 20 years.

13 Conclusion

The last century allowed for the discovery of isolated and linear physiopathology models that paved the way for mass medications (e.g., antibiotics in infectiology). It also brought about a revolution for mechanical assistance for acute and emergency illnesses (e.g., controlled mechanical ventilation, extracorporeal circulation, orthopedic surgery). The downside of these considerable forms of progress is that we now face a growing number of people living with slowly evolving chronic diseases (e.g., cancer, diabetes, neurodegenerative disease, cardiovascular syndrome). These diseases are complex, sneaky, and unpredictable due to their multifactorial origins (biological, behavioral, psychological, social, economic, and environmental). Biotechnology treatments fail to cure these diseases, which affect an exponentially increasing number of people as a result of the aging of the world's population. These treatments consume three-quarters of a state's healthcare costs, without achieving a total recovery, and, in some cases, causing more harm than good – as is attested by the withdrawal of drugs from the market.

The vertical organization of a health system centered on curative objectives and expectations of a miracle treatment for the entire population is no longer adapted to the health challenges that await us in this new century. Faced with these shortcomings, a multitude of health solutions, sometimes inspired by traditional medicine, sometimes available with new technologies, are emerging. These solutions are based on an epigenetic approach, a resolutely preventive approach, and on digital traceability. They carry with them aspirations to give meaning to individual care, to health, and to an individual's life journey. They take up a considerable place in society and they are called NPIs.

Health authorities are barely aware of the value of NPIs, which are popular with the majority of citizens and with avant-garde health professionals. An ecosystem is born.

> **Key Points**
>
> - The siloed organization of a health system centered on curative objectives for the entire population is no longer adapted to the health challenges that await us in this new century (e.g., chronic disease, aging).
> - Faced with these shortcomings, a multitude of health solutions, sometimes inspired by traditional medicine, sometimes available with new technologies, are emerging. These solutions are based on an epigenetic approach, a resolutely preventive approach, and on digital traceability. They carry with them aspirations to give meaning to individual care, to health, and to an individual's life journey. They take up a considerable place in society, and they are called NPIs.
> - Health authorities are barely aware of the value of NPIs, which are popular with the majority of citizens and with avant-garde health professionals. An ecosystem is born.

References

Bacon, S. L., Lavoie, K. L., Ninot, G., Czajkowski, S., Freedland, K. E., Michie, S., Montgomery, P., Powell, L. H., & Spring, B. (2015). An international perspective on improving the quality and potential of behavioral clinical trials. *Current Cardiovascular Risk Reports, 9*(427), 2–6. https://doi.org/10.1007/s12170-014-0427-0.

Barnett, K., Mercer, S. W., Norbury, M., Watt, G., Wyke, S., & Guthrie, B. (2012). Epidemiology of multimorbidity and implications for health care, research, and medical education: A cross-sectional study. *Lancet, 380*(9836), 37–43. https://doi.org/10.1016/S0140-6736(12)60240-2.

Bernard, C. (1865). *Introduction à l'étude de la médecine expérimentale*. Paris: J.B. Baillière et Fils.

Boutron, I., Moher, D., Altman, D. G., Schulz, K. F., & Ravaud, P. (2008). Extending the CONSORT statement to randomized trials of nonpharmacologic treatment: Explanation and elaboration. *Annals of Internal Medicine, 148*, 295–309. https://doi.org/10.7326/0003-4819-148-4-200802190-00008.

Bowling, A. (2001). *Measuring disease* (2nd ed.). Philadelphia: Open University Press.

CAMbrella. (2012). A pan-European research network for complementary and alternative medicine. In *CAMbrella Report*.

Carey, N. (2012). *The epigenetics revolution: How modern biology is rewriting our understanding of genetics, disease, and inheritance*. New York: Columbia University Press.

Cohen, J. T., Neumann, P. J., & Weinstein, M. C. (2008). Does preventive care save money? Health economics and the presidential candidates. *New England Journal of Medicine, 358*(7), 661–663. https://doi.org/10.1056/NEJMp0708558.

Corroller-Soriano, A. L., Malavolti, L., & Mermilliod, C. (2008). *La Vie deux ans après le diagnostic de cancer. Ministère du travail, des relations sociales et de la solidarité*. Paris: Documentation Française.

Critchley, J. A., & Capewell, S. (2003). Mortality risk reduction associated with smoking cessation in patients with coronary heart disease: A systematic review. *The Journal of the American Medical Association, 290*(1), 86–97. https://doi.org/10.1001/jama.290.1.86.

Curtis, J. R., Martin, D. P., & Martin, T. R. (1997). Patient-assessed health outcomes in chronic lung disease: What are they, how do they help us, and where do we go from here? *American Journal of Respiratory and Critical Care Medicine, 156*, 1032–1039. https://doi.org/10.1164/ajrccm.156.4.97-02011.

De Ridder, D., Geenen, R., Kuijer, R., & van Middendorp, H. (2008). Psychological adjustment to chronic disease. *Lancet, 372*, 246–255. https://doi.org/10.1016/S0140-6736(08)61078-8.

Domingo-Salvany, A., Lamarca, R., Ferrer, M., Garcia-Aymerich, J., Alonso, J., Félez, M., Khalaf, A., Marrades, R. M., Monsó, E., Serra-Batlles, J., & Antó, J. M. (2002). Health-related quality of life and mortality in male patients with chronic obstructive pulmonary disease. *American Journal of Respiratory and Critical Care Medicine, 166*, 680–685. https://doi.org/10.1164/rccm.2112043.

Ernst, E. (2011). Fatalities after CAM: An overview. *British Journal of General Practice, 61*(587), 404–405. https://doi.org/10.3399/bjgp11X578070.

Fabbri, L. M., Luppi, F., Beghe, B., & Rabe, K. E. (2008). Complex chronic comorbidities of COPD. *European Respiratory Journal, 31*(1), 204–212. https://doi.org/10.1183/09031936.00114307.

Flores, M., Glusman, G., Brogaard, K., Price, N. D., & Hood, L. (2013). P4 medicine: How systems medicine will transform the healthcare sector and society. *Personalized Medicine, 10*(6), 565–576. https://doi.org/10.2217/pme.13.57.

Frey, U., & Suki, B. (2008). Complexity of chronic asthma and chronic obstructive pulmonary disease: Implications for risk assessment, and disease progression and control. *Lancet, 20*(372), 1088–1099. https://doi.org/10.1016/S0140-6736(08)61450-6.

GBD 2017 Disease and Injury Incidence and Prevalence Collaborators. (2018). Global, regional, and national incidence, prevalence, and years lived with disability for 354 diseases and injuries for 195 countries and territories, 1990-2017: A systematic analysis for the global burden of disease study 2017. *Lancet, 392*(10159), 1789–1858. https://doi.org/10.1016/ S0140-6736(18)32279-7.

Global Wellness Institute. (2018). *Global wellness economy monitor*. Miami: GWI.

Gøtzsche, P. (2013). *Deadly medicines and organised crime: How big pharma has corrupted healthcare*. London: Radcliffe publishing Ltd..

Greenier, K. D., Kernis, M. H., McNamara, C. W., Waschull, S. B., Berry, A. J., Herlocker, C. E., & Abend, T. A. (1999). Individual differences in reactivity to daily events: Examining the roles of stability and level of self-esteem. *Journal of Personality, 67*, 185–208. https://doi. org/10.1111/1467-6494.00052.

Griffiths, T. L., Burr, M. L., Campbell, I. A., Lewis-Jenkins, V., Mullins, J., Shiels, K., Turner-Lawlor, P. J., Payne, N., Newcombe, R. G., Ionescu, A. A., Thomas, J., & Tunbridge, J. (2000). Results at 1 year of outpatient multidisciplinary pulmonary rehabilitation: A randomised controlled trial. *Lancet, 355*(9201), 362–368. https://doi.org/10.1016/s0140-6736(99)07042-7.

Khaw, K. T., Wareham, N., Bingham, S., Welch, A., Luben, R., & Day, N. (2008). Combined impact of health behaviours and mortality in men and women: The EPIC-Norfolk prospective population study. *PLoS Medicine, 5*, e12. https://doi.org/10.1371/journal.pmed.0050012.

Kuchenbaecker, K. B., Hopper, J. L., Barnes, D. R., et al. (2017). Risks of breast, ovarian, and contralateral breast cancer for BRCA1 and BRCA2 mutation carriers. *The Journal of the American Medical Association, 317*(23), 2402–2416. https://doi.org/10.1001/jama.2017.7112.

Li, K., Hüsing, A., & Kaaks, R. (2014). Lifestyle risk factors and residual life expectancy at age 40: A German cohort study. *BMC Medicine, 12*, 59. https://doi.org/10.1186/1741-7015-12-59.

Li, Y., Pan, A., Wang, D. D., Liu, X., Dhana, K., Franco, O. H., Kaptoge, S., Di Angelantonio, E., Stampfer, M., Willett, W. C., & Hu, F. B. (2018). Impact of healthy lifestyle factors on life expectancies in the US population. *Circulation, 138*(4), 345–355. https://doi org/10.1161/ circulationaha.117.032047.

Lim, S. S., Vos, T., Flaxman, A. D., et al. (2012). A comparative risk assessment of burden of disease and injury attributable to 67 risk factors and risk factor clusters in 21 regions, 1990-2010: A systematic analysis for the global burden of disease study 2010. *Lancet, 380*(9859), 2224–2260. https://doi.org/10.1016/S0140-6736(12)61766-8.

Lindholm, M. E., Marabita, F., Gomez-Cabrero, D., Rundqvist, H., Ekström, T. J., Tegnér, J., & Sundberg, C. J. (2014). An integrative analysis reveals coordinated reprogramming of the epigenome and the transcriptome in human skeletal muscle after training. *Epigenetics, 9*(12), 1557–1569. https://doi.org/10.4161/15592294.2014.982445.

MacKeigan, L. D., & Pathak, D. S. (1992). Overview of health-related quality-of-life measures. *American Journal of Hospital Pharmacy, 49*, 2236–2245.

Manuel, D. G., Perez, R., Sanmartin, C., Taljaard, M., Hennessy, D., Wilson, K., Tanuseputro, P., Manson, H., Bennett, C., Tuna, M., Fisher, S., & Rosella, L. C. (2016). Measuring burden of unhealthy behaviours using a multivariable predictive approach: Life expectancy lost in Canada attributable to smoking, alcohol, physical inactivity, and diet. *PLoS Medicine, 13*, e1002082. https://doi.org/10.1371/journal.pmed.1002082.

Maurer, J., Rebbapragada, V., Borson, S., Goldstein, R., Kunik, M., Yohannes, A., & Hanania, N. A. (2008). Anxiety and depression in COPD: Current understanding, unanswered questions, and research needs. *Chest, 134*(4S), 43S–56S. https://doi.org/10.1378/chest.08-0342.

Mozaffarian, D., Wilson, P. W., & Kannel, W. B. (2008). Beyond established and novel risk factors: Lifestyle risk factors for cardiovascular disease. *Circulation, 117*(23), 3031–3038. https://doi. org/10.1161/circlutationaha.107.738732.

Ninot, G. (2013). *Démontrer l'efficacité des interventions non médicamenteuses: Question de points de vue*. Montpellier: Presses Universitaires de La Méditerranée.

Ninot, G. (2020). *301 reviews on NPIs published in 2019*. Blog en santé. http://blogensante.fr/ en/2020/03/02/301-revues-sur-les-inm-de-litterature-publiees-en-2019/

Ninot, G., Delignieres, D., & Varray, A. (2010). Stability of physical self: Examining the role of chronic obstructive pulmonary disease. *European Review of Applied Psychology, 60*, 35–40. https://doi.org/10.1016/j.erap.2009.07.001.

O'Doherty, M. G., Cairns, K., O'Neill, V., Lamrock, F., Jørgensen, T., Brenner, H., Schöttker, B., Wilsgaard, T., Siganos, G., Kuulasmaa, K., Boffetta, P., Trichopoulou, A., & Kee, F. (2016). Effect of major lifestyle risk factors, independent and jointly, on life expectancy with and without cardiovascular disease: Results from the consortium on health and ageing network of cohorts in Europe and the United States (CHANCES). *European Journal of Epidemiology, 31*, 455–468. https://doi.org/10.1007/s10654-015-0112-8.

Prince, M., Bryce, R., & Ferri, C. (2011). *The benefits of early diagnosis and intervention: World Alzheimer Report, Alzheimer's Disease*. Washington: Alzheimer's Disease International (ADI).

Queneau, P., & de Bourguignon, C. (2017). *Sauver le médecin généraliste*. Paris: Éditions Odile Jacob.

strategy, F. i.-m. (2007). Plan pour l'amélioration de la qualité de vie des personnes atteintes de maladies chroniques. In *Paris*.

Tamakoshi, A., Tamakoshi, K., Lin, Y., Yagyu, K., & Kikuchi, S. (2009). Healthy lifestyle and preventable death: Findings from the Japan collaborative cohort (JACC) study. *Preventive Medicine, 48*, 486–492. https://doi.org/10.1016/j.ypmed.2009.02.017.

Vogeli, C., Shields, A. E., Lee, T. A., Gibson, T. B., Marder, W. D., Weiss, K. B., & Blumenthal, D. (2007). Multiple chronic conditions: Prevalence, health consequences, and implications for quality, care management, and costs. *Journal of General Internal Medicine, 22*, 391–395. https://doi.org/10.1007/s11606-007-0322-1.

Ware, J. E. (2003). Conceptualization and measurement of health-related quality of life: Comments on an evolving field. *Archives of Physical Medicine and Rehabilitation, 84*, 43–51. https://doi.org/10.1053/apmr.2003.50246.

WHO. (2003). *Adherence to long-term therapies: Evidence for action*. Geneva: WHO.

WHO. (2005). *Preparing a health care workforce for the 21st century: The challenge of chronic condition*. Geneva: WHO.

WHO. (2006). *Preventing chronic diseases: A vital investment: WHO global report*. Geneva: WHO.

WHO. (2008). *2008–2013 action plan for the global strategy for the prevention and control of noncommunicable diseases*. Geneva: WHO.

WHO. (2013a). *Global action plan for the prevention and control of noncommunicable diseases 2013–2020*. Geneva: WHO.

WHO. (2013b). *WHO traditional medicine strategy: 2014–2023*. Geneva: WHO.

WHO. (2014). *Global status report on noncommunicable diseases*. Geneva: WHO.

WHO. (2015). *Global action plan on antimicrobial resistance*. Geneva: WHO.

WHOQOL Group. (1994). Development of the WHOQOL: Rationale and current status. *International Journal of Mental Health, 56*, 23–24.

Wiesener, S., Kalkenberg, T., Hegyi, G., Hök, J., Roberti di Sarsina, P., & Fønnebø, V. (2012). Legal status and regulation of complementary and alternative medicine in Europe. *Forschende Komplementmedizin, 19*(S2), 29–36. https://doi.org/10.1159/000343125.

Xu, Y., Su, D., Zhu, L., Zhang, S., Ma, S., Wu, K., Yuan, Q., & Lin, N. (2018). S-allylcysteine suppresses ovarian cancer cell proliferation by DNA methylation through DNMT1. *Journal of Ovarian Research, 11*(1), 39. https://doi.org/10.1186/s13048-018-0412-1.

Ziegelstein, R. C. (2017). Personomics: The missing link in the evolution from precision medicine to personalized medicine. *Journal of Personalized Medicine, 7*(4), 11. https://doi.org/10.3390/jpm7040011.

Chapter 3
Mechanisms Involved in Non-pharmacological Interventions

1 Introduction

Both qualitative and quantitative research, including empirical and experimental studies, are making progress in discovering the mechanisms that lead to the health benefits provided by non-pharmacological interventions (NPIs). These mechanisms are characterized by their multiplicity, and they most often function in a simultaneous and systematic manner. While on one hand, researchers must avoid the pretension of explaining all NPI effects on humans in terms of their biological, psychosocial, and historical singularity, they must also be wary of falling into the trap of dismissing everything that is inexplicable or mysterious as a major danger and an open door to paranormalism. As such, this chapter provides an overview of the mechanisms that are involved when NPIs are implemented.

2 The Natural History of Illness and Spontaneous Remission

Before focusing on the mechanisms of NPIs, it is important to discuss the natural course of diseases and symptoms. This is a critical point to consider as it is often overlooked in consideration of the therapeutic process and can be easily mistaken as a result of a treatment when it is a natural attribute of the human body. For example, subjective symptoms like pain tend to be cyclical, which means that patients who only see a practitioner when their symptoms flare up are likely to see improvements over time, regardless of what the practitioner does or does not do. In these cases, patients simply misinterpret the improvement as an effect of treatment.

The Natural Course of a Disease or a Symptom
Even without external intervention, time can heal. The human body contains internal resources with which it fights against pathogens, repairs the damage caused by accidents or diseases, and relieves pain as much as possible.

2.1   Even without Action, Time Heals

Self-healing or spontaneous remission exists and can occur regardless of the treatment chosen. Over the course of 200,000 years, the bodies of *Homo sapiens* have learned to defend themselves from viruses such as the flu, and bacteria such as *Streptococcus pneumoniae*, and to heal themselves from injuries such as wounds, twisted ligaments, and muscular tears. Some NPIs such as physiotherapy treatments and psychotherapies mobilize all of the body's internal resources to fight a disease.

The Self-Healing Observations of Physician Andrew Weil
In a famous book in the United States, physician Andrew Weil (1995) reported on the clinical observations of dozens of cases of spontaneous recovery. He also recounts his findings based on trips he made over the course of 15 years to study experts in traditional medicine, shamans, medicine men, and healers. He found, for example, that Chinese medicine focuses on strengthening the body's internal defenses by increasing the efficiency and activity of cells. According to Weil, the body is constantly repairing all parts of itself, from cells to tissues. This "restorative system" depends on the coordinated interaction of the stimulating and inhibiting factors that influence the proliferation and development of cells. This is not just a reaction to an injury, but rather a fundamental principle of a normal healthy tissue. This system should be celebrated as it is responsible for the moments we spend in good health. The body wants to be healthy and will seek to recover itself in the event of an imbalance. As Hippocrates once said, healing is a natural power. The body is a whole and all of its parts are linked; it works as a holistic system with built-in restorative powers that can be called upon. His exploratory and innovative work led Weil to recommend strategies that help the body restore and maintain its own balance: deep breathing, sleep, physical activity, diet, avoiding exposure to toxic substances, changing one's outlook on disease, and maintaining a good therapeutic relationship with one's doctor. These strategies help to support the body's natural "repairing," "regeneration," and spontaneous healing capacities as a means of preventing or complementing biomedical treatments.

3   Biological Mechanisms

It is impossible to create an inventory of all of the biological mechanisms behind the benefits of NPIs in a single book. Nevertheless, by using different biological scales we can organize some of these mechanisms into a logical frame.

3.1   At the Organ and Tissue Level

With the hope of immortality, humans have always sought to counter the effects of aging and the defects it engenders in the body's functional parts. For example, we can recall the ancient Greek myth of Prometheus which demonstrates the knowledge of the body's potential to repair itself. In the myth Prometheus, who is tied to a rock for having given fire to men, must watch an eagle devour his liver every day. Each day his liver regenerates so that his punishment can be eternally repeated. The ancient Greeks' choice of the liver over any other organ for the myth illustrates that they knew the liver has significant regenerative abilities. When as much as three quarters of the liver is amputated, it regains its size in only four months.

Even though human organ development ends at around 20 years of age, they are continuously regenerated through the cell cycle process in which thousands of billions of cells are created, die, and are replaced at variable frequencies according to their function. This process has significant effects on the body and varies considerably between individuals. For example, a New Zealand study of 1037 individuals who were 38 years old showed that the biological age of their bodies ranged from 30 to 60 years (Belsky et al. 2005).

Taken together, the body's regenerative process and its variability from person to person explain one aspect of the mechanism of NPIs. NPIs have shown the potential to improve the function and structure of different body organs including muscles, such as physical NPIs like physical therapies and manual therapies; joints, such as physical NPIs like physiotherapy and balneotherapy; bones, such as elemental NPIs like mineral and botanical preparations; the heart, such as nutritional NPIs like diets and supplements; the lungs, such as psychological NPIs like health-education programs promoting healthy lifestyle changes; and the brain, such as digital NPIs like virtual reality and video-game-based therapies that develop cognitive functions.

Moreover, an exercise program that reduces body fat also improves muscle mass and delays joint arthrosis. As a result, not only is the muscle improved in quantity through the addition of new muscular fibers, it also has improved in quality through the amelioration of its capillarization, the increased specialization of its fibers according to the efforts required, an improved binding of the actin and myosin strands, an increased neuromotor recruitment of its fibers, and increased tendon elasticity. Other examples of NPI mechanisms on the level of organ and tissue regeneration include NPIs targeting smoking cessation. Even though the rib cage stiffens from age 30, stem cells in the lungs can create bronchioles or alveoli. In addition, 1 year after smoking cessation, the risk of myocardial infarction is halved, and after 10 years, the risk of lung cancer is halved. Thus, we can see how different NPIs can work together with the body's regenerative capacities in order to help promote its natural self-repairing processes. For example, many NPIs bring about lifestyle changes that improve the body's functioning and repair by minimizing exposure to toxins such as alcohol or conditions such as obesity, which make the body less effective. NPIs stimulate the body's physiological adaptation and regeneration resulting in improvements in its functioning.

A High-Quality Methodological Study on Acupuncture
The German Acupuncture Trials for chronic low back pain (GERAC) randomized 1162 patients with a history of chronic low back pain persisting for at least 8 years into three groups: the first group received 10- to 30-minute acupuncture sessions twice a week with 14–20 needles; the second group received acupuncture only on "sham" points ("insensitive" or "non-point" areas in which the acupuncture needles are placed on areas of the skin that are outside of the meridian paths); and the third group received a conventional treatment. Of the three groups, the acupuncture group demonstrated the most improvement in pain and mobility scores after 6 months. The control group showed modest improvement, and the "sham" group showed a marked improvement that was greater than the control group, but less significant than the acupuncture group. The unexpected effectiveness of "non-point" acupuncture raises questions about the overall mechanisms of acupuncture, and there is little convincing experimental data beyond studies on acupuncture in the treatment of pain. Local anesthesia at the point of insertion of acupuncture needles blocks the effects, suggesting that the effects of acupuncture are related to innervation. However, acupuncture also increases local blood flow and is associated with a release of adenosine (the analgesic effects of which are known) at the needle insertion site, as was demonstrated in a recent publication in *Nature Neurosciences* (Goldman et al. 2010). There is evidence suggesting that acupuncture is associated with the release of endogenous opioids in the brainstem, subcortical, and limbic structures. For example, functional MRI in humans shows the immediate and prolonged effects of acupuncture on the limbic system (both in the somatosensory and in the affective functions involved in painful processes). In addition, studies with PET scans show that acupuncture increases the potential for binding to μ receptors for several days, while other studies indicate that it also affects the release of stress hormones such as ACTH (Li et al. 2008).

3.2   At the Cellular Level

Studies have shown the capacity of the salamander to regenerate its members. Even though humans do not have the same capacity as the salamander, the zebrafish, or even a common plant for total regeneration, recent studies indicate that this capacity could exist, or at least that it could be stimulated. Regenerative capacity is mainly controlled by the cells of the body that can reprogram themselves to replace damaged cells, tissues or organs. Some of these so-called "stem" cells are generated by the bone marrow and can circulate in the body. Other stem cells are generated by the tissues themselves, such as stem cells in the hair follicles that ensure hair growth throughout life. Regardless of their points of origin, these cells have the potential to transform themselves in order to repair and grow all kinds of tissue in the body. This potential is called cell differentiation, and it gives rise to many hopes for expanding the human body's potential, especially since stem cells have also been identified in the central nervous system (brain) and peripheral nervous system (bone marrow and nerves). Even if human regeneration may not seem immediately obvious, there is a common example that everyone has experienced – scarring. However, scarring shows that regeneration is only partial; this mechanism does not allow for an identical reproduction. For example, you only have to see the regenerated skin of burn victims to see that it is only a partial reproduction of the elasticity of the original skin tissue. In order to advance in this domain, we still need to understand the reason for this insufficiency.

Chromosome and Telomerase
Three researchers, Elizabeth Blackburn, Carol Greider, and Jack Szostak, 2009 Nobel Prize winners in medicine, have shown that the shortening of telomeres, over the course of cell divisions, causes cells to age. An enzyme, telomerase, affects its length and plays a role in slowing aging and cell cancerization. This encourages researchers to understand how to influence this enzyme.

Beyond stem cells, in vitro or in vivo studies show that exercise programs stimulate immunity, anti-inflammatory processes, anti-oxidation, and metabolism (INSERM 2019). Analgesic (pain reliever), anti-asthenic (anti-fatigue), and anti-stress effects also have been observed in various mechanistic studies. Effects on sleep and on digestive function are observed. Exercise programs improve cell functioning, such as mitochondria. Muscles release myokines.

3.3   At the Molecular Level

Epigenetics involves the study of temporary biochemical changes in the genome that are caused by diverse environmental factors. One such change is methylation, in which a methyl group is added to or removed from a base in the DNA molecule without affecting the original DNA sequence (Carey 2012). As an illustration, if genes can be likened to the hardware of the cell, epigenetic mechanisms can be considered as their software (Grazioli et al. 2017). Multiple studies indicate that stress transforms the human epigenome (Ziegelstein 2017). This means that stress induces epigenetic marks, which can be thought of as small accessories impacting the DNA by modifying the cell's reading of the genetic code. These changes can have various consequences; for example, one of the epigenetic effects of stress is the acceleration of aging. Additionally, epigenetic marks due to stress also tend to increase the risks of chronic diseases and psychopathologies. However, since marks can be deposited on DNA, they can also be removed or their locations can be changed. Furthermore, it appears as though NPIs play a role in this process, which can be considered as one of the underlying mechanisms of their effectiveness. A recent study published in *Brain, Behavior and Immunity* reinforces this idea (Chaix et al. 2020). Researchers assessed the impact of an eight-hour day of intensive meditation on the epigenome of 17 experienced practitioners who had been meditating for at least 3 years and for at least 30 minutes a day. The results were compared with a control group of 17 people who had never meditated, but who carried out various leisure activities during an eight-hour day. Within the meditation group more than 60 methylation sites (a form of epigenetic modification) had undergone significant changes after a day of intensive meditation. The sites concerned are mainly associated with the metabolism and aging of immune cells. Thus, this study suggests that epigenetic marks can be altered by very short-term therapies.

4   Neurological Mechanisms

Recent advances in neuroscience, notably due to brain imaging and microbiology, are bringing to light various essential properties of the brain, such as plasticity. Studies are beginning to show how this sophisticated organ copes with and structurally changes in response to environmental stimuli. They are also giving us a better understanding of how the brain is connected to the whole body. As such, it represents a growing and valuable domain in which to consider the mechanisms of NPIs.

4.1   Automatic Processes

Cognitive training programs specifically stimulate executive functions to enable optimized learning. Synaptic connections are at work to consolidate the various learnings and make them automatic processes requiring less control and less attention. For example, learning to drive a car is not that easy. It requires an extensive amount of attention and control. After a relatively short period of training sessions, you can drive without realizing that you are driving; your brain will be busy thinking about something else during the trip. This is possible because actions initially controlled by higher brain functions have been automated and no longer require as much vigilance. In the event of an alert, the controlled processes instantly take over.

4.2   Neural Plasticity

Whereas 20 years ago it was claimed that areas of the brain were specialized and that cognitive functions were irreversibly lost in the event of an injury or doomed to decline with aging, recent studies in neural plasticity indicate that the brains of patients suffering from cerebrovascular accidents (stroke) can reorganize their neural pathways according to the hindrances of stresses and stimuli (Bolte Taylor 2009), regardless of their age. We now know that new synaptic connections are created and that the hemispheres can be reshaped. We also know that there are exchanges that occur between the cerebral hemispheres.

Some neurosurgeons, like Hugues Duffau (2013), challenge structuralist and fixist theories of the brain. New learning methods, actions to mobilize cognitive resources, and re-education strategies are being developed for the purpose of demonstrating and developing the potential of the brain's plasticity. Over time, these rehabilitation methods become fully-fledged NPIs that act in accordance with the brain's plasticity and adaptability. Some of these methods are purely neurocognitive; others mobilize sensory pathways (using music, for example); others use bodily biofeedback; and others use digital technologies to compensate for deficits through applications (daily reminders of tasks via smartphone for Alzheimer's disease).

4.3   Neurogenesis

Similar to the case of neural plasticity, 20 years ago it was thought that we were born with brains containing a fixed number of neurons that could only decrease with advancing age. However, the discovery of stem cells in the brain and the principles of adult neurogenesis show that new neurons are formed in the brain. The challenge is to be able to create the conditions that facilitate neurogenesis with relevant associated NPIs (e.g., virtual reality training program).

4.4   Somatic Nervous System

The body remains permanently connected to the environment through the somatic nervous system and its integrator, the parietal cortex. All our sensory organs and sensory systems – auditory, visual, olfactive, motricity, touch, and even the skin – are constantly receiving numerous signals from the environment. NPIs "play" with these biosensors, restoring them when they are in faulty operation, to compensate, to balance. For example, a study of Sahaja yoga practitioners shows a greater volume of gray matter (cell body of neurons and connections) than that of non-practitioners (Hernandez et al. 2016).

4.5   Neuroendocrine Activity

A psychological NPI such as psychotherapy can moderate stress via the hypothalamic-pituitary-adrenal axis. Over time, the moderation of these neuroendocrine dynamics can optimize various physiological processes that are involved in oxidative metabolism, DNA repair, gene expression, and the production of growth factors (Antoni et al. 2006). For example, if a tumor is present, neuroendocrine factors can regulate the activity of the proteases, angiogenic factors, cytokines, and adhesion molecules involved in tumor progression.

4.6   Sympathetic and Parasympathetic Systems

Research increasingly suggests that NPIs can regulate stress by acting on the sympathetic and parasympathetic nervous systems. For example, at the HeartMath Institute work is underway on advancing our understanding of cardiac coherence, which constitutes a state of optimal synergy between respiration, heart rate, and psychophysiological coherence. Heart rate variability (HRV) biofeedback training with wearable devices is associated with a large reduction in self-reported stress and anxiety (Goessl et al. 2017).

NPIs influence the sympathetic system associated with two neurotransmitters, norepinephrine and adrenaline, which prepare the body for action in the face of stress. They generate a so-called fight or flight response, which leads to the dilation of the bronchi, the acceleration of cardiac and respiratory activities, increased blood pressure, dilated pupils, increased sweating, and decreased digestive activity.

Furthermore, certain NPIs can also work in parallel on the parasympathetic system, which prepares the body for relaxation via the neurotransmitter acetylcholine. This induces a general slowing down of the vital functions of the organism. Commonly known as the "rest and digest" mechanism, the parasympathetic system favors digestive functions and the sexual appetite. It also slows down the heart rate and respiratory activity and lowers blood pressure, all of which are associated with a state of relaxation shown to be beneficial for the restoration of the organism. Some digital NPIs have been developed for this purpose such as music-therapy devices.

4.7   Emotional Regulation

Some psychological NPIs can directly influence our moods and emotional states. These changes occurred in the absence of detectable increases in prefrontal control systems. They provoke moments of pleasure and joy that leave an imprint on the brain and deflect negative thoughts. They decrease significantly stress markers such as cortisol. For example, emotion regulation using mindfulness training was associated with reductions in reported pain and negative affect, with reduced amygdala responses to negative images and with reduced heat-evoked responses in medial and lateral pain systems (Kober et al. 2019). Mindful acceptance significantly reduced activity in a distributed a priori neurologic signature that is sensitive and specific to experimentally induced pain. Thus, a momentary mindful acceptance regulates emotional intensity by changing initial appraisals of the affective significance of stimuli, which has consequences for clinical treatment of pain and emotion (Kober et al. 2019).

Without necessarily speaking of happiness, the body becomes a vector of pleasure and not of suffering. Many psychosomatic studies and qualitative surveys confirm these positive impacts on emotions and emotional regulation (Buckworth et al. 2013).

4.8   Neurotransmission

NPIs mobilize the mechanisms linked to the neural reward circuit. Many NPIs such as art therapies, electronic games, and various body methods allow their participants to experience moments of pleasure and laughter. As a result, the practice of these NPIs can augment the production of endorphins such as dopamine and serotonin. For example, studies indicate that yoga increases levels of gamma-aminobutyric acid (GABA), which is the most common neurotransmitter in the brain as well as an important facilitator of relaxation and sleep that also has been indicated to mitigate symptoms of depression (Streeter et al. 2010, 2020). Another NPI, cognitive behavioral therapy (CBT), has been shown to modulate levels of glutamate in the brain, which is the substance from which GABA is made (Abdallah et al. 2014). Therefore, the practice of this NPI could also cause neurophysiological changes in the cortex that could improve the transmission of GABA (Radhu et al. 2012).

Oxytocin is a hormone that is secreted by the paraventricular and supraoptical nuclei of the hypothalamus and excreted by the posterior pituitary gland. It plays an important role in human social bonding, facilitating sympathy, generosity, and sociability by making us more sensitive to and benevolent toward others. Increased oxytocin levels have been indicated to make us feel more sympathetic to our loved ones. Each time we meet our loved ones, it activates mirror neurons in our brains. As a result, through the process of positive synchronization, our emotions and emotional states synchronize with those of our loved ones. NPIs strongly use this process.

4.9   The Second Brain

The gut is now being seen as constituting a "second brain," and researchers are increasingly explaining how this organ acts as a major player in our health (Enders 2017). Beyond the gut's capacity to directly analyze our food and drink intake in addition to its digestive functions, it remains in permanent connection with the brain largely via the vagus nerve.

It is worth underscoring that while the intestine is subjected to most or all microbial and parasitic attacks on the body, the brain remains protected by its bony housing, membranous covering, tissue shell, and neurological and blood filters. Meanwhile, the gut continuously reports what is going on in the body to the brain, simultaneously acting as a modulator or aggravator of stress.

Nutritional NPIs, manual therapies, and tailored physical activity programs have been shown to be particularly efficient in influencing this "gut weather" directly related to health status.

5   Psychological Processes

Humans have long been amazed by the mysterious influence that the mind has on the body. Over the past decade, this field has been the subject of renewed interest for both scientists and clinicians. The use of psychological interventions in the treatment of various health problems is becoming increasingly legitimized by the fact that many causes of disease have psychological origins, such as stress, anxiety, self-esteem disorders, or chronic fatigue. We are also increasingly seeing the negative effects of the behavioral origins of many diseases such as smoking and sedentary lifestyles. Several scientific disciplines are devoted to focusing on the underlying psychological processes of disease, in particular psychosomatics, positive psychology, health psychology, and clinical psychology. They indicate that fostering the individual's sense of control is a key process leading to the path of health and well-being and that it is underpinned by important socio-cognitive processes such as the development self-confidence, causal attribution, motivation, and self-efficacy.

5.1 Cognition

Research indicates that selection and information processes also play a role in optimizing the mind-body connection and, by extension, our overall well-being. Various NPIs such as different psychosomatic practices, physical activities, and digital therapies train patients to focus on important and useful information and to ignore spurious information and irrelevant ruminations. They encourage users to remain centered in the present by focusing on the task at hand and eliminating futile thoughts. For example, who has never experienced feelings of health, clarity, and present-centeredness at the end of a long run, a bike ride, a trying hike up a mountain, or a feat of swimming? Certain NPIs also help to reduce procrastination processes in which you spend more time planning to do things than actually doing them. For example, research on mindfulness meditation indicates its usefulness in reducing procrastination and developing strategies for following through. This practice helps individuals to focus their energy and attention on the present, which is beneficial to problem solving and overall stress reduction. This focused attention is becoming increasingly important as our lives are ever more punctuated by the chronic stress and ill health effects resulting from the disturbances generated by electronic devices that are supposed to make our lives easier but that keep us in a state of constant alertness in which we jump at the slightest beep or screen lighting up.

Learning to remain in a state of focused attention optimizes our information collection and processing. Certain NPIs such as disease management programs can enhance the process of relevant information collection, can reduce biases in analysis, and can decrease distortion in information expression. As a result, decisions become clearer as they are made with increased speed and conviction. This renders actions that are more fluid and precise. Other NPIs develop the use of visualization, positive thinking, and focusing techniques in order to minimize intrusive thoughts, thereby allowing patients to distract themselves from their symptoms and to forget about their disease for a bit.

Beck's Cognitive Model of Depression
Cognition is central to the treatment of depression since emotions, thoughts, and behaviors are all mediated by cognitive processes. Cognitive behavioral therapy (CBT) is an NPI that follows the tenets of Beck's cognitive model of depression, which suggests that problematic experiences early in life can make the individual vulnerable to depression as a result of their development of negative core beliefs about self, others, and the world. Later in life, these beliefs can be triggered by stressful events. During episodes of depression, information processing is affected, leading to biased and negative interpretations of interpersonal experiences that increase the individual's sense of isolation and reduce their levels of activity. CBT is based on three fundamental assumptions: cognitive activity affects behavior; cognitive activity can be monitored and altered; and that desired behavior change can be affected through cognitive change. The therapist works collaboratively with the patient in order to help them identify, explore, and modify their relationships with negative thinking, behavior, and depressed moods. This is achieved by teaching the patient to identify and monitor the intensity of their different moods, to recognize the thoughts and behaviors that contribute to these moods, and to learn strategies for evaluating and challenging unhelpful thoughts and engaging in behaviors that contribute to improved moods. Cognitive change in depression is achieved by targeting negative, automatic thoughts (immediate and plausible negative thoughts based on faulty schema rather than on reflective reasoning), and by using techniques such as thought catching, reality testing, task assignment, and the development of alternative strategies. Behavioral experiments are also used to help patients re-evaluate their underlying beliefs and assumptions.

5.2   Metacognition

A plethora of research demonstrates how small the proportion of information that reaches our consciousness is compared to our total brain activity (Bolte Taylor 2009). These underlying dynamics, such as embodied cognition, strongly influence our decision making, learning, and behavior. For example, a study explored the effects that two interventions had on the development of the sitting motor skill and problem-solving capacity of children with developmental delays and/or cerebral palsy ranging in age from 8 to 34 months (Harbourne and Berger 2019). One of the interventions emphasized motor-based problem solving while the other focused on advancing motor skills through assistance in order to attain optimal movement patterns. Participants in the problem-solving group showed significant gains in their Early Problem Solving for Infants scores over the participants in the optimal movement patterns group. This indicates that cognitive embodiments are a major source of decisions and behaviors including tremors, sweating, nonverbal gestures, body positions, movements, smells, noises, tastes, touching, sounds, voice tones, looking, choice of shoes, choice of clothing, and even hairstyles. Informal brain capture is considerable; thought is so restrictive.

5.3 Psycho-Emotional Processes

Certain NPIs – such as many of the psychological, physical, and psychosomatic practices – generate positive emotions and pleasure. This satisfaction acts as a release from both acute and chronic stress. NPIs involving relaxation and meditation techniques inspired from Buddhist or Hindu practices are reputed to create paths toward inner peace. They aim to restore the individual's connection with their whole body. Currently, mindfulness practices are booming throughout the Western world. Some of these NPIs, such as the *mindfulness-based stress reduction* (MBSR) technique created by Kabat-Zinn (2018), have good track records of evidence for their efficacy.

Csikszentmihalyi and Csikszentmihalyi (2006) explain the benefits of these types of methods with the notion of flow. "Flow" is a state of peace with oneself, in which the individual is able to "let go" of their fixation on self and judgments. In this state everything becomes fluid and more evident. It generates feelings of fullness and an opening to *momentum*. The individual is able to feel united with their body, in the *Hic et Nunc* – the here and now.

Another psycho-emotional process that is too rarely mentioned in medical and personal development books concerns sexuality. It is rarely mentioned within the patient care process and is treated as if it should no longer exist in the case of disease. Therapists have started to think about this topic, without going to extremes like Wilhelm Reich, a sexologist and disciple of Sigmund Freud. As a result, various psychologically based NPIs are beginning to develop around this topic as we become increasingly aware of the importance of maintaining and restoring sexuality as part of the patient-healing process. Not only is sexuality a vital impulse, it is also vector of well-being, including among the elderly and in patient populations.

5.4 Psychosocial Levels

On a personal level, it cannot be denied that the efficacy of many NPIs is directly linked to an individual's psychosocial context and its influence on the type and level of their expectations. For example, in one study in which patients were asked about their opinion of acupuncture, the results showed that those with high levels of expectation had a much more positive vision of the technique than those with more limited expectations (Linde et al. 2007). NPIs tend to be centered on the individual and to promote personalized and progressive goals that can enable participant success. This is because individual improvements toward one's goals can be more easily identified and observed – in some cases, they can even be photographed or filmed – and can be attributed to the NPI. This helps people feel more competent, more motivated, and more optimistic about continuing to advance in order to achieve their goals. As they progress, they look to the future with a greater sense of control. These psychological benefits can be extrapolated to the health field. For example, an increased sense of control can help patients feel more able to manage their treatments. They can build confidence and gain autonomy, which empowers them so that they feel they can act on their disease. Many NPIs activate socio-cognitive processes of perceived control, self-efficacy (Bandura 1997), and intrinsic motivation in order to consolidate an individual's sense of empowerment. Cognitive reassessment with positive expectations and self-efficacy then supports the adoption of healthy behaviors and feeling of control (Bandura 1997).

On an interpersonal level, the effectiveness of NPIs is often reliant on the support of relatives and caregivers as a means of giving patients additional motivation. For example, people who no longer want to dress or leave their homes sometimes can be persuaded to do so in order to please their children and their loved ones. In the case of NPIs that are practiced in group settings, such as in group-based psychotherapies or collective psychosomatic or physical practices, patients can be exposed to others, which may help them to say that they cannot complain about other more seriously ill patients. Furthermore, within the context of positive psychology interventions, such as those studied by Tal Ben-Shahar of Harvard University, the benevolent gaze of loved ones often gives patients the extra energy needed to cope with their disease. Along these lines, patient associations can propose NPIs to strengthen these interpersonal processes.

Therapeutic Alliance
The patients who cope the best with their disease are those who are able to make their healthcare teams an ally rather than an opponent. Additionally, dialogue with other patients coping with the same disease is often essential to achieve optimal disease management.

On a social level, beliefs and stereotypes directly influence health. They forge convictions and can consolidate our determination to act, to hope, or to resign ourselves. In return, these ideas about our own state of health have a profound impact on us. An observational study that followed 660 50-year-old individuals over a 20-year period showed that people who had positive perceptions of their aging live, on average, 7-and-a-half years longer than those who perceive it negatively (Levy et al. 2002). The researchers are beginning to understand that imagining and convincing ourselves that we can and are being healed is an important means of accomplishing it. Seeking out the support of fellow patients, patient experts, or patient testimonies in NPI education programs can provide comfort, courage, and a positive structure for patient identity. Some patients use social networks to campaign for their cause – for example, against cancer, against tobacco. Others find a new social role as patient experts in order to support people suffering from the same illness.

5.5   Meta-Psychology

Since the dawn of time humans have held onto belief in order to face an uncertain and chaotic future. Nowadays, some of the spiritually inspired NPIs that discuss crucial issues (life, death, love, family) are beginning to be accepted in medical structures beyond palliative care units – as long as they do not disrupt medical routines. Seligman and Csikszentmihalyi's (2000) notion of positive psychology, "to begin to catalyze a change in the focus of psychology from preoccupation only with repairing the worst things in life to also building positive qualities" (p. 5), was determinant in this regard. Many NPIs of this type have been proven to foster happiness, subjective well-being, optimism, and self-determination. This is because many of them focus on promoting key concepts such as acceptance, resilience, and sense of coherence.

> **Acceptance**
> The acceptance of a disease is a continual psychological process. Saying that it only takes a stay in a health facility to learn how to manage a chronic disease is a mistake. Saying that a bariatric surgery is enough to lose weight is an aberration. Learning to live with a disease is a long-term, self-learning process. It involves unlearning bad lifestyle habits and changing behaviors that are harmful for health. Through acceptance, the patient assumes a part of the responsibility for changes in their health status. They become an actor as they learn to take better care of themselves. It's a process that takes time. With success, the patient sees their illness differently, giving it its place, all of its place, but nothing but its place.

Resilience is the ability to make find strength in a difficult life event such as a disease. For example, it is common to meet people who have been positively transformed by a disease. They are animated by a deep inner peace, and when they discuss their illnesses they do not express a defensive attitude or aggressiveness toward their existence with a disease. Many say that they used their illness as an opportunity for change. They learned to live with it, to tame it, to minimize its consequences, even if they never forget it. Accepting a disease is not a passive process. It reflects a proactive approach that is never completely accomplished. Some NPIs help advance the individual's resilience process by facilitating acceptance and helping patients to make sense of and relate their symptoms or illness within their own life story. One telling example of this can be found in Anna Halprin's use of dance and movement as a means of coming to terms with and overcoming her own cancer.

Sense of Coherence

"global orientation that expresses the extent to which one has a pervasive, enduring though dynamic feeling of confidence that one's internal and external environments are predictable and that there is a high probability that things will work out as well as can reasonably be expected" (Antonovsky 1979, p. 123).

Aaron Antonovsky (1979) thought that a sense of coherence is an influence on an individual's health status. He depicted it as a concept shaped through the repeated experience of the availability of and successful coping with general resistance resources. Many NPIs address the contextual dimensions of a salutogenic model (Mittelmark et al. 2020); they are rooted in the social determinants of health and confront its political determinants. Examples include participatory health literacy programs and self-management programs, which promote the motivational dimensions of a sense of coherence within social support, comprehensibility, manageability, and meaningfulness.

6   The Placebo Effect

In the nineteenth century, Armand Trousseau devoted several studies to the placebo phenomenon in the Hôtel Dieu Hospital of Paris. The placebo effect carries elements within it that are fascinating, supernatural, and inexplicable. It challenges the curiosity of clinicians, researchers, experimentalists, and methodologists; from yesterday to nowadays, including in the field of NPIs. The placebo effect operates in humans regardless of age, sociocultural level, personality, and lifestyle (Benedetti et al. 2011; Kaptchuk 1998; Sullivan 1993).

The pivotal article on this effect was published in 1955 by Henry K. Beecher in the *Journal of the American Medical Association.* Beecher noted, on the basis of scientific and medical facts, that the placebo effect – defined as the difference between the patient's condition before and after taking a placebo product – occurs, on average, 35.2% in the case of any therapy.

A placebo product, by definition, cannot be effective without the placebo effect. An active product, on the other hand, can be effective because of the placebo effect. If the intake of aspirin calms a patient within 10 min, this is due to the placebo effect. It cannot be the result of the active principle of the molecule, since the physiological action time of the active principle is approximately 1 h.

The work of Fabrizio Benedetti's team has since greatly advanced knowledge on this mechanism and its involvement during the administration of any prescribed treatment.

The placebo effect occurs 30%, on average, for any therapy and can reach as much as 70% in the treatment of migraine and depression. A review on the efficacy

> **Defining a Placebo and the Placebo Effect**
> The term placebo is a Latin word meaning: "I will please." In medicine, it implies: "I will please anyone who asks me to prescribe treatment."
> *"A placebo is not the inert substance alone, but rather its administration within a set of sensory and social stimuli that tell the patient that a beneficial treatment is being given"* (Benedetti et al. 2011, p. 339). The placebo is a therapeutic measure of zero or low intrinsic efficacy, with no logical relation to the disease, but acting as if the subject thinks he is receiving active treatment.
> A real placebo effect is a psychobiological phenomenon occurring in the patient's brain after the administration of an inert substance, or of a sham physical treatment such as sham surgery, along with verbal suggestions (or any other cue) of clinical benefit (Benedetti et al. 2011, p. 339).

of antidepressant drugs in patients suffering from a major depressive disorder showed that the placebo effect accounted for 51% of the therapeutic effect, while the effect of the active molecule and the effect of spontaneous remission accounted for only 25% and 24%, respectively (Kirsch and Sapirstein 1998). The placebo effect could be even more important for preventive actions.

6.1 A Biological Phenomenon

The placebo effect corresponds to a psychobiological phenomenon that occurs in the patient's brain following the administration of an inert treatment that is accompanied by verbal suggestions of clinical benefit. It results in an improvement in the patient's condition, corresponding to the positive difference between the observed therapeutic result and the foreseeable therapeutic effect according to biological models (Dispenza 2014). The contribution of the brain's dopaminergic system to the placebo effect is no longer in doubt (Benedetti et al. 2011). A placebo intervention stimulates the body's production of endorphins producing an analgesic effect. As a result, patients experience less pain, but improvements explained by only a biological mechanism would be too simple.

6.2 A Psychosocial Phenomenon

The placebo effect also results from psychosocial mechanisms that act simultaneously, sometimes sequentially, and not always with the same intensity:

- Anticipation: the patient prepares their body and mind in anticipation of coping with an event. In the case of pain, the patient activates their primary sensorimotor cortex, insular cortex, and anterior cingulate cortex. There are several mechanisms through which the expectation of a future event may affect different physiological functions. Expectations of a positive outcome prepare the body in anticipation of an event, in order to better cope with it; as such, it offers a clear evolutionary advantage. A cognitive reassessment of the appropriate conduct is done to inhibit negative expectations. In the opposite case, expectations of a negative outcome such as pain may result in the amplification of pain, and several brain regions – such as the anterior cingulate cortex, the prefrontal cortex, the insula, and the hippocampus – have been found to be activated during the anticipation of pain (Benedetti et al. 2011).

– Anxiety: a placebo can modulate emotions (Petrovic et al. 2005). The endogenous opioid system is activated. Attention, particularly in the prefrontal cortex, plays a role in focusing on or avoiding focusing on the pain stimulus. In so doing, the placebo effect decreases patient anxiety.

– Reward: the mesolimbic dopaminergic system is activated when the patient expects a placebo effect.

– Individual learning: patients who have regularly experienced pain (trauma) such as headaches more easily associate a characteristic of a placebo with a phenomenon of pain reduction. For example, a person may have been conditioned over time by the color of a pill. This repeated experiential learning stimulates the immune system, the endocrine system, and the neuromuscular system.

– Social learning: patients can learn the placebo effect through observation and imitation. A placebo has a positive impact in patients who are informed that they are taking a placebo as opposed to a control group who does not take it and who knows it.

– A relationship of trust: the relationship between the patient and the doctor is essential to potentiate a placebo effect. If the doctor treats the patient with conviction and optimism, rather than with skepticism, the patient will be more confident and optimistic. For example, one study compared the intensity of pain when morphine is delivered by a machine or explained by a doctor. The pain

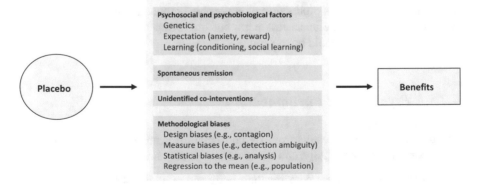

Fig. 3.1 Sub-mechanisms of placebo effect. (Adapted by permission from Springer Nature: Benedetti et al. 2011)

decreases more quickly when the injection is given by a doctor rather than by a machine (Colloca et al. 2004).

6.3   A Holistic Phenomenon

Thus, the placebo effect is an intrinsic part of any solution that a human gives to another human with the aim of improving the latter's health. It is a complex process that is at once biological, psychological, relational, and cultural (Fig. 3.1). Beliefs and learning amplify its effects. Both the therapeutic context and the therapist-patient alliance are major components in the stimulation of the placebo effect because they create trust and mutual efficiency expectations.

To conclude, the administration of an NPI provokes a complex cascade of biochemical events that are activated by several social stimuli. Such events will inevitably contribute to the responses observed. Treatments are not given to a robot, but rather they are introduced into a complex biochemical environment that varies according to the patient's cognitive/affective state and their previous exposure to other treatments (Benedetti et al. 2011).

7   Nursing Effect

Verbal empathy and listening are the pillars upon which a caregiver or an actor of prevention using an NPI must rely. NPI users must be afforded optimal conditions of appeasement, confidence, and security, which will allow them to bring buried traumas to the surface, to draw out untapped resources from themselves, and to enable them to surpass themselves for their own benefit. Anesthesiologists have compared the sensation of pain experienced by pregnant women when an anesthetic is injected. The control group was prepared for the injection with the common warning that, "*You will feel like an intense bee sting. This is the most unpleasant part of the intervention.*" The experimental group was informed differently: "*We will give you a local anesthetic that will numb you, so that you feel good during the operation.*" In close relation with the discussion of the placebo effect, this study showed that the use of soothing words had a significant impact on the patient's sensation of pain and the degree of discomfort during the surgical intervention (Varelmann et al. 2010).

Defining the Nursing Effect
The nursing effect is a relational, material, and environmental ambience that is created by a health professional and in which the patient is immersed. This "bath of kindness" encourages the patient to take care of themselves in response to the efforts that are made by the health professionals to treat them. This "bath of attention" has a significant impact on the evolution of the patient's health condition. The attention that is paid to the patient and the way in which they are addressed creates a relationship of trust. During this privileged, intimate, and singular moment everything matters: the tone of voice and words used; the health professional's look, posture, and gestures; and their breathing, smell, and clothing.

In other words, beyond medical procedures, healthcare professionals' nonverbal gestures, posture, looks, and attitude are all important determinants of the nursing effect. These nonverbal cues act as testimonies of the caregiver's real attentiveness, availability, authenticity, and kindness. A patient receiving positive scrutiny from medical personnel may expect more therapeutic benefits because of the special attention received.

The quality and appearance of medical materials also contribute to the nursing effect. The safer, more reliable, more ergonomic, and more comfortable these materials are, the more they act as a testimony to the engagement and professionalism of the care team.

The environment is the last contributor to nursing effect. Hospital, health center, and senior residences are starting to pay more and more attention to the quality of their services (e.g., food, TV), interior design, and architecture. Beyond the functional aspects (accessibility, for example) and safety aspects (bathroom, for example), the idea is to put the patient in a setting with the best possible conditions. The white, austere rooms and sanitized, lifeless odors that characterized the hospitals and retirement homes from the end of the last century are gradually disappearing from the healthcare landscape.

Now health facility managers are working with ergonomists and architects to rethink the places of care and collective life in order to turn them into spaces for living, exchanging, and sharing. Collective spaces like dining rooms and entrance halls are being revised. Increasingly, the exteriors of these facilities are also beginning to be used as spaces for practicing NPIs, for making landscaped trails, therapeutic gardens, and meeting spaces.

8   Multimodal Effects

Exercise involves several mechanisms on the biological level. Researchers have shown that during exercise there is an increase in the whole-body oxygen uptake, heart rate and cardiac production, ventilation, arterial partial pressure of oxygen, and hemoglobin saturation. Skeletal muscles increase the adenosine triphosphate turnover; glycogenolysis; glucose uptake; lipolysis; free fatty acid uptake; oxygen utilization; carbon dioxide and heart production; blood capillarization; capillary recruitment; release of biologically active molecules called myokines with autocrine, paracrine, and endocrine effects; hypertrophy; and mitochondrial biogenesis. The metabolic system expands the liver's glucose production, mainly from glycogenolysis, but also from gluconeogenesis. There is an increase of adipose tissue lipolysis and free fatty acid mobilization. The central nervous system regulates the physiological response to exercise through somatic and autonomous systems. Cerebral blood flow and oxygen supply are better maintained. "The study of exercise biology shows that the need to integrate observations from genes, molecules, and cells in a physiological context has never been greater" (Hawley et al. 2014, p. 747).

Obviously, it is useful for researchers to try to understand what is happening in the "black box" through experimental studies isolating each individual action mechanism that is mobilized by an NPI. However, this analytical approach with its origins in targeted biology remains limited in that it cannot provide a picture of the effects that introducing a psychological, physical, digital, or nutritional program has on a patient's everyday life. A 3-month intensive exercise program mobilizes biopsychosocial mechanisms that act simultaneously from the molecular to the psychosocial scales. Its impacts will be observable and measurable for multiple indicators on multiple scales (Vina et al. 2012). A patient does not live in the isolated, ideal world that is reflected in a laboratory experiment or an in vitro study. In "real life" the physiological and psychological mechanisms that are activated by an NPI act simultaneously, working and interacting together in the improvement of a patient's health status.

Due to their mobilization of parallel and interacting mechanisms, the effects of NPIs cannot be considered as cumulative or additional; rather, they are exponential products of systemic interactions. When several NPIs are implemented at the same time, such as through the combination of an exercise program with a dietary program, more biotherapeutic mechanisms are mobilized and more effects become potentialized.

Simultaneous Antidepressant Mechanisms Mobilized by an Exercise Program

NPIs, such as supervised exercise programs, activate several biological mechanisms and psychosocial processes simultaneously. Their antidepressant effects do not depend on a single mechanism.

Physiological studies have found that these programs stimulate a better supply of oxygen to the central nervous system, improvements in cerebral blood flow, adaptive responses to the production of reactive oxygen, and enhanced oxidative stress resilience (de Sousa et al. 2017). They also have been indicated to reduce the circulating levels of several inflammatory biomarkers through the release of muscle-derived myokines (Brown et al. 2015).

Neurobiological studies have found that these programs promote changes in brain morphology through adaptive improvements in cerebral blood flow (Maass et al. 2015). This facilitates neurogenesis in the hippocampus and hippocampal neuroplasticity (Yau et al. 2011), having a direct influence on the release of cortisol and serotonin as well as on the stimulation of the endorphin pathways and facilitation of the reward circuit (Greer and Trivedi 2009).

Neurocognitive studies have found that these programs modify the individual's level of conscious awareness, in particular by increasing their body awareness (Mehling et al. 2011). This improves executive functions (planning, coordination, focus, learning) and minimizes the distraction and diversion caused by negative and intrusive thoughts (Craft et al. 2007).

Psychosocial studies have found that these programs increase the individual's sense of physical self-perception and self-esteem due to improvements in muscle mass and a reduction in body fat (Annesi and Porter 2015). They also improve the individual's sense of control, which improves behavioral strategies for coping with stressful situations. In so doing, aversive cycles preventing the individual from pursuing future goals or initiating tasks break down, which allows for increases in self-efficacy through improved physical abilities and the learning of new skills (White et al. 2009). By extension, this increases the individual's sense of pleasure and improves their overall social participation.

Faced with the multiplicity and simultaneity of the action mechanisms mobilized by NPIs, more and more researchers are calling for a paradigm shift and a change in the model to allow for the conceptualization of integrated theories. The overlapping in the activated processes of NPIs require their consideration as a "systemic molecule" that is too complex for the reductionism of the Cartesian mind that seeks to identify a single explanation for a health status change. Rather, the effects of NPIs more closely resemble the multimodal chains in an open environment that account for the human complexity that is well-described by Edgar Morin (2008).

Some researchers have developed integrated theories that account for changes in health behaviors over time. For example, James O. Prochaska and his team have

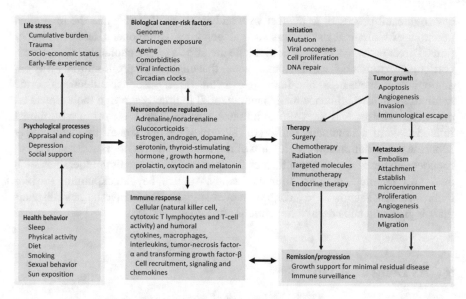

Fig. 3.2 Integrated model of bio-behavioral influences on cancer pathogenesis through neuroendocrine pathways. (Adapted by permission from Springer Nature: Antoni et al. 2006)

developed a circular model in five successive steps that explains the lasting change in a health behavior such as smoking cessation. In this transtheoretical model (Prochaska and Prochaska 2011), the individual passes from pre-contemplation, to contemplation, to preparation, into action and maintenance. Within this model a relapse phase is possible. The theory behind this model has made it possible to develop specific NPIs that offer behavioral techniques relevant to each phase of the process to smokers who wish to quit. For example, imagine telling a smoker who has no intention of quitting smoking (pre-contemplation stage) that tobacco is bad for their health. Not only would this not encourage them to quit smoking, it also could have the opposite effect and increase their smoking. Over the course of decades of implementing NPIs targeted at smoking cessation, practitioners have observed that one behavior change can lead to another. Helping a patient to quit smoking thus allows them to shift their focus to increasing physical activity, then to reducing alcohol consumption, and then to changing to healthy eating. Considering this virtuous behavioral spiral led the Prochaska team to realize that regardless of what constituted the initial behavioral change, the subsequent behavioral changes it engendered could serve as a shield to prevent relapse (Prochaska and Prochaska 2011).

In the same vein, Antoni et al. (2006) proposed an integrative model in which bio-behavioral factors are shown to influence multiple aspects of tumorigenesis through their impact on neuroendocrine function. Their research demonstrates that bio-behavioral factors such as life stress, psychological processes, and health behaviors influence tumor-related processes through the neuroendocrine regulation of hormones, including adrenaline, noradrenaline, and glucocorticoids (Fig. 3.2). Central control of peripheral endocrine functions also allows behavioral, social, and environmental processes to interact with and systemically modulate the malignant potential of biological risk factors such as genomes, carcinogens, and viral infections. Direct pathways of influence include the effects of catecholamines and glucocorticoids on tumor cell expressions of genes that control cell proliferation, invasion, angiogenesis, metastasis, and immune evasion. Stress-responsive neuroendocrine mediators can also influence malignant potential indirectly through their effects on oncogenic viruses and the cellular immune system. These pleiotropic hormonal influences induce a mutually reinforcing system of cellular signals that collectively suppress the initiation and progression of malignant cell growth. Furthermore, neuroendocrine deregulation can influence the response to conventional therapies such as surgery, chemotherapy, and immunotherapy. As such, beyond explaining the biobehavioral risk factors for cancer, this model suggests innovative targets for pharmacological and behavioral interventions (Antoni et al. 2006).

9   As Long as It Works

Mechanistic research on NPI effectiveness has progressed over the past 10 years. This growing body of research highlights NPI mechanisms that mobilize the body's resources, modify its structures, compensate for its deficiencies, adapt its functions, and promote its regeneration.

Nevertheless, the fact remains that many healings seem to continue to escape human rationality. Physics has two explanations for light, a wave or a particle. On a scale even smaller than light, the quantum scale, cause-and-effect relationships are even less obvious (Ball 2017). Science never stops. Perhaps as it continues to progress, science will discover new rationalities that are able to explain the mechanisms behind NPI benefits. Maybe artificial intelligence will help?

As it currently stands, everything is still not seen, everything is still not understood, and everything still cannot be explained. Despite this lack of understanding, for certain patients in certain contexts, certain inexplicable NPIs work, such as healers' methods and medicine-men's remedies. As Cabrera and collaborators have reminded us since 2015, a lack of proof of effectiveness is not proof of a lack of effectiveness. From a methodological viewpoint, the evidence of a significant effect that a clinical trial may demonstrate seems to depend more often on the evaluation protocol chosen than the actual results obtained (Cohen-Mansfield 2013).

As such, if mechanistic studies remain necessary for us to better understand what is happening in the body and the mind of a patient and how it is affected by the singular relationship between a patient and a practitioner (or a team), interventional studies and cohort studies are needed in order to verify in which patients NPIs work, how they work, and what the benefits and risks are.

Going beyond the realities of their systemic mechanisms, the ever-present role of the placebo effect, and the potential for spontaneous self-healing, the main thing is that there is empirical evidence that attests to the risk-benefit ratio of various NPIs.

This is essential in order to recommend their precise use, to set up monitoring devices, to better train practitioners, and to properly inform users.

NPI Calibration

Optimizing solutions, finding the right dose and the right combinations, and avoiding the creation of myths, these are the keys to advancing our understanding and use of NPIs. The explanatory mechanisms behind them do not really matter, as long as a specific NPI decreases pain, relieves symptoms, and contributes to the well-being of a patient. NPIs require patience, persistence, and effort in order to generate real benefits. Furthermore, these benefits are rarely immediate beyond individual perception.

Within the field of human health research, a number of authors such as Hawe and Potvin (2009) encourage the development of a science of solutions as an addition to the science of problems. This pragmatic and constructivist approach encourages the development of an applied and transdisciplinary approach that complements fundamental and analytical approaches.

10 Conclusion

NPIs cure few diseases; however, they prevent, treat, and forewarn many. They also mobilize several bio-behavioral mechanisms to act simultaneously and with cascading effects. Functionally, some NPIs mobilize the body's residual resources, modify epigenetic tags, boost metabolism, activate the immune system, strengthen anti-inflammatory activity, stimulate body-mind connections, reduce stress, and restore confidence and positive expectations. In so doing, they potentiate the processes of self-healing. Structurally, some NPIs regenerate cells, make tissues, transform the intestinal microbiota, and repackage organs. In so doing, they modify epigenetic configurations. These effects require time and the proactive participation of NPI users. Above all, perhaps the great strength of NPIs is that they restore the human capacity to hope for a cure and to feel a sense of control, even if it is only partial. On the patient level, it seems as though the explanatory mechanisms behind NPIs do not really matter as long as they decrease the pain, relieve the symptoms, and help contribute to overall well-being.

Key Points

Contrary to drugs and surgeries with specific targets and a unique action mechanism, NPIs simultaneously mobilize a cascade of several biopsychosocial mechanisms. Functionally, NPIs mobilize the body's residual resources, influence the epigenome, boost metabolism, activate the immune system, strengthen anti-inflammatory activity, stimulate body-mind connections, reduce stress, and restore patient confidence and hope. Structurally, NPIs regenerate cells, make tissues, transform the intestinal microbiota, repackage organs, and in so doing, modify epigenetic configurations. These effects require both patients' time and effort, which increases the influence of the placebo effect. By helping to restore hope, self-confidence, and meaning to patients' lives, NPIs potentiate self-healing processes. The main goal or focus for future research in this domain should be optimizing NPI solutions and finding the right doses and combinations for their implementation.

References

Abdallah, C. G., Niciu, M. J., Fenton, L. R., Fasula, M. K., Jiang, L., Black, A., Rothman, D. L., Mason, G. F., & Sanacora, G. (2014). Decreased occipital cortical glutamate levels in response to successful cognitive-behavioral therapy and pharmacotherapy for major depressive disorder. *Psychotherapy and Psychosomatics, 83*(5), 298–307. https://doi.org/10.1159/000361078.

Annesi, J. J., & Porter, K. J. (2015). Reciprocal effects of exercise and nutrition treatment-induced weight loss with improved body image and physical self-concept. *Behavioral Medicine, 41*(1), 18–24. https://doi.org/10.1080/08964289.2013.856284.

Antoni, M. H., Lutgendorf, S. K., Cole, S. W., Dhabhar, F. S., Sephton, S. E., McDonald, P. G., Stefanek, M., & Sood, A. K. (2006). The influence of bio-behavioural factors on tumour biology: Pathways and mechanisms. *Nature Reviews Cancer, 6*(3), 240–248. https://doi.org/10.1038/nrc1820.

Antonovsky, A. (1979). *Health, stress and coping.* San Francisco: Jossey-Bass.

Ball, P. (2017). A world without cause and effect. *Nature, 546*, 590–592.

Bandura, A. (1997). *Self-efficacy: The exercise of control.* New York: Cambridge University Press.

Beecher, H. K. (1955). The powerful placebo. *The Journal of the American Medical Association, 159*, 1602–1606.

Belsky, D. W., Caspi, A., Houts, R., Cohen, H. J., Corcoran, D. L., Danese, A., Harrington, H. L., Israel, S., Levine, M. E., Schaefer, J. D., Sugden, K., Williams, B., Yashin, A. I., Poulton, R., & Moffitt, T. E. (2005). Quantification of biological aging in young adults. *Proceedings of the National Academy of Sciences of the United States of America, 112*(30), E4104–E4110. https://doi.org/10.1073/pnas.1506264112.

Benedetti, F., Carlino, E., & Pollo, A. (2011). How placebos change the patient's brain. *Neuropsychopharmacology Reviews, 36*, 339–354. https://doi.org/10.1038/npp.2010.81.

Bolte Taylor, J. (2009). *My stroke of insight: A brain Scientist's personal journey.* New York: Pinguin Group.

Brown, W. M., Davison, G. W., McClean, C. M., & Murphy, M. H. (2015). A systematic review of the acute effects of exercise on immune and inflammatory indices in untrained adults. *Sports Medicine Open, 1*(1), 35. https://doi.org/10.1186/s40798-015-0032-x.

Buckworth, J., Dishman, R. K., O'Connor, P. J., & Tomporowski, P. (2013). *Exercise psychology.* Champaign: Human Kinetics.

Cabrera, E., Sutcliffe, C., Verbeek, H., Saks, K., Soto-Martin, M., Meyer, G., Leino-Kilpi, H., Karlsson, S., & Zabelegui, A. (2015). Non-pharmacological interventions as a best practice strategy in people with dementia living in nursing homes, a systematic review. *European Geriatric Medicine, 6*, 134–150. https://doi.org/10.1016/j.eurger.2014.06.003.

Carey, N. (2012). *The epigenetics revolution: How modern biology is rewriting our understanding of genetics, disease and inheritance*. New York: Icon Books.

Chaix, R., Fagny, M., Cosin-Tomás, M., Alvarez-López, M., Lemeeef, L., Regnaulte, B., Davidson, R. J., Lutz, A., & Kalimanh, P. (2020). Differential DNA methylation in experienced meditators after an intensive day of mindfulness-based practice: Implications for immune-related pathways. *Brain, Behavior, and Immunity, 84*, 36–44. https://doi.org/10.1016/j.bbi.2019.11.003.

Cohen-Mansfield, J. (2013). Nonpharmacologic treatment of behavioral disorders in dementia. *Current Treatment Options in Neurology, 15*, 765–785. https://doi.org/10.1007/s11940-013-0257-2.

Colloca, L., Lopiano, L., Lanotte, M., & Benedetti, F. (2004). Overt versus covert treatment for pain, anxiety, and Parkinson's disease. *Lancet Neurology, 3*, 679–684. https://doi.org/10.1016/S1474-4422(04)00908-1.

Craft, L. L., Freund, K. M., Culpepper, L., & Perna, F. M. (2007). Intervention study of exercise for depressive symptoms in women. *Journal of Womens Health, 16*(10), 1499–1509. https://doi.org/10.1089/jwh.2007.0483.

Csikszentmihalyi, M., & Csikszentmihalyi, I. S. (2006). *A life worth living: Contributions to positive psychology*. New York: Oxford University Press.

de Sousa, C. V., Sales, M. M., Rosa, T. S., Lewis, J. E., de Andrade, R. V., & Simões, H. G. (2017). The antioxidant effect of exercise: A systematic review and meta-analysis. *Sports Medicine, 47*(2), 277–293. https://doi.org/10.1007/s40279-016-0566-1.

Dispenza, J. (2014). *You are the placebo: Making your mind matter*. New York: Hay House.

Duffau, H. (2013). *Diffuse low-grade gliomas in adults: Natural history, interaction with the brain, and new individualized therapeutic strategies*. New York: Springer.

Enders, G. (2017). *Gut: The inside story of our body's most underrated organ*. London: Scribe Publications.

Goessl, V. C., Curtiss, J. E., & Hofmann, S. G. (2017). The effect of heart rate variability biofeedback training on stress and anxiety: A meta-analysis. *Psychological Medicine, 47*(15), 2578–2586. https://doi.org/10.1017/S0033291717001003.

Goldman, N., Chen, M., Fujita, T., Xu, Q., Peng, W., Liu, W., Jensen, T. K., Pei, Y., Wang, F., Han, X., Chen, J. F., Schnermann, J., Takano, T., Bekar, L., Tieu, K., & Nedergaard, M. (2010). Adenosine A1 receptors mediate local anti-nociceptive effects of acupuncture. *Nature Neuroscience, 13*, 883–888. https://doi.org/10.1038/nn.2562.

Grazioli, E., Dimauro, I., Mercatelli, N., Wang, G., Pitsiladis, Y., Di Luigi, L., & Caporossi, D. (2017). Physical activity in the prevention of human diseases: Role of epigenetic modifications. *BMC Genomics, 18*(S8), 802. https://doi.org/10.1186/s12864-017-4193-5.

Greer, T. L., & Trivedi, M. H. (2009). Exercise in the treatment of depression. *Current Psychiatry Reports, 11*(6), 466–472.

Harbourne, R. T., & Berger, S. E. (2019). Embodied cognition in practice: Exploring effects of a motor-based problem-solving intervention. *Physical Therapy, 99*(6), 786–796. https://doi.org/10.1093/ptj/pzz031.

Hawe, P., & Potvin, L. (2009). What is population health intervention research? *Canadian Journal of Public Health, 100*, I8–I14.

Hawley, J. A., Hargreaves, M., Joyner, M. J., & Zierath, J. R. (2014). Integrative biology of exercise. *Cell, 159*, 738–749. https://doi.org/10.1016/j.cell.2014.10.029.

Hernandez, S. E., Suero, J., Barros, A., Gonzalez-Mora, J. L., & Rubia, K. (2016). Increased grey matter associated with long-term Sahaja yoga meditation: A voxel-based morphometry study. *PLoS One, 11*(3), e0150757. https://doi.org/10.1371/journal.pone.0150757.

INSERM. (2019). *Physical activity: Prevention and treatment of chronic diseases*. Paris: INSERM.

Kabat-Zinn, J. (2018). *Meditation is not what you think*. New York: Hachette Book Group.

Kaptchuk, T. J. (1998). Powerful placebo: The dark side of the randomised controlled trial. *Lancet,* *351,* 1722–1725. https://doi.org/10.1016/S0140-6736(97)10111-8.

Kirsch, I., & Sapirstein, G. (1998). Listening to Prozac but hearing placebo: A meta-analysis of antidepressant medication. *Prevention and Treatment, 1,* 1–16. https://doi.org/10.1037/1522-3736.1.1.12a.

Kober, H., Buhle, J., Weber, J., Ochsner, K. N., & Wager, T. D. (2019). Let it be: Mindful acceptance down-regulates pain and negative emotion. *Social Cognitive and Affective Neuroscience, 14*(11), 1147–1158. https://doi.org/10.1093/scan/nsz104.

Levy, B. R., Slade, M. D., Kunkel, S. R., & Kasl, S. V. (2002). Longevity increased by positive self-perceptions of aging. *Journal of Personality and Social Psychology, 83*(2), 261–270. https://doi.org/10.1037//0022-3514.83.2.261.

Li, A., Lao, L., Wang, Y., Xin, J., Ren, K., Berman, B. M., et al. (2008). Electroacupuncture activates corticotrophin-releasing hormone-containing neurons in the paraventricular nucleus of the hypothalamus to alleviate edema in a rat model of inflammation. *BMC Complementary and Alternative Medicine, 8,* 20. https://doi.org/10.1186/1472-6882-8-20.

Linde, K., Witt, C. M., Streng, A., Weidenhammer, W., Wagenpfeil, S., Brinkhaus, B., Willich, S. N., & Melchart, D. (2007). The impact of patient expectations on outcomes in four randomized controlled trials of acupuncture in patients with chronic pain. *Pain, 128,* 264–271. https://doi.org/10.1016/j.pain.2006.12.006.

Maass, A., Düzel, S., Goerke, M., Becke, A., Sobieray, U., Neumann, K., Lövden, M., Lindenberger, U., Bäckman, L., Braun-Dullaeus, R., Ahrens, D., Heinze, H. J., Müller, N. G., & Düzel, E. (2015). Vascular hippocampal plasticity after aerobic exercise in older adults. *Molecular Psychiatry, 20*(5), 585–593. https://doi.org/10.1038/mp.2014.114.

Mehling, W. E., Wrubel, J., Daubenmier, J. J., Price, C. J., Kerr, C. E., Silow, T., Gopisetty, V., & Stewart, A. L. (2011). Body awareness: A phenomenological inquiry into the common ground of mind-body therapies. *Philosophy, Ethics, and Humanity in Medicine, 6*(6), 1–12. https://doi.org/10.1186/1747-5341-6-6.

Mittelmark, M. B., Sagy, S., Eriksson, M., Bauer, G. F., Pelikan, J. M., Lindström, B., & Espne, G. A. (2020). *The handbook of salutogenesis.* New York: Springer.

Morin, E. (2008). *On complexity.* New York: Hampton Press.

Petrovic, P., Dietrich, T., Fransson, P., Andersson, J., Carlsson, K., & Ingvar, M. (2005). Placebo in emotional processing-induced expectations of anxiety relief activate a generalized modulatory network. *Neuron, 46,* 957–969. https://doi.org/10.1016/j.neuron.2005.05.023.

Prochaska, J. J., & Prochaska, J. O. (2011). A review of multiple health behavior change interventions for primary prevention. *American Journal of Lifestyle Medicine, 5*(3). https://doi.org/10.1177/1559827610391883.

Radhu, N., Daskamakis, Z. J., Guglietti, C. L., Farzan, F., Barr, M. S., Arpin-Cribbie, C. A., Fitzgerald, P. B., & Ritvo, P. (2012). Cognitive behavioral therapy-related increases in cortical inhibition in problematic perfectionists. *Brain Stimulation, 5*(1), 44–54. https://doi.org/10.1016/j.brs.2011.01.006.

Seligman, M. E. P., & Csikszentmihalyi, M. (2000). Positive psychology: An introduction. *American Psychologist, 55,* 5–14. https://doi.org/10.1037//0003-066x.55.1.5.

Streeter, C. C., Gerbarg, P. L., Brown, R. P., Scott, T. M., Nielsen, G. H., Owen, L., Sakai, O., Sneider, J. T., Nyer, M. B., & Silveri, M. M. (2020). Thalamic Gamma Aminobutyric Acid Level changes in major depressive disorder after a 12-week Iyengar yoga and coherent breathing intervention. *Journal of Alternative and Complementary Medicine, 26*(3), 190–197. https://doi.org/10.1089/acm.2019.0234.

Streeter, C. C., Whitfield, T. H., Owen, L., Rein, T., Karri, S. R., Yakhkind, A., Perlmutter, R., Prescot, A., Renshaw, P. F., Ciraulo, D. A., & Jensen, E. (2010). Effects of yoga versus walking on mood, anxiety, and brain GABA levels: A randomized controlled MRS study. *Journal of Alternative and Complementary Medicine, 16*(11), 1145–1152. https://doi.org/10.1089/acm.2010.0007.

Sullivan, M. D. (1993). Placebo controls and epistemic control in orthodox medicine. *Journal of Medicine and Philosophy, 18*, 213–231. https://doi.org/10.1093/jmp/18.2.213.

Varelmann, D., Pancaro, C., Cappiello, E., & Camann, W. R. (2010). Nocebo-induced hyperalgesia during local anesthetic injection. *Anesthesia and Analgesia, 110*, 868–870. https://doi.org/10.1213/ANE.0b013e3181cc5727.

Vina, J., Sanchis-Gomar, F., Martinez-Bello, V., & Gomez-Cabrera, M. C. (2012). Exercise acts as a drug: The pharmacological benefits of exercise. *British Journal of Pharmacology, 167*(1), 1–12. https://doi.org/10.1111/j.1476-5381.2012.01970.x.

Weil, A. (1995). *Spontaneous healing: How to discover and enhance your body's natural ability to maintain and heal itself*. New York: Alfred A. Knopf.

White, K., Kendrick, T., & Yardley, L. (2009). Change in self-esteem, self-efficacy and the mood dimensions of depression as potential mediators of the physical activity and depression relationship: Exploring the temporal relation of change. *Mental Health and Physical Activity, 2*(1), 44–52. https://doi.org/10.1016/j.mhpa.2009.03.001.

Yau, S. Y., Lau, B. W. M., Tong, J. B., Wong, R., Ching, Y. P., Qiu, G., Tang, S. W., Lee, T. M. C., & So, K. F. (2011). Hippocampal neurogenesis and dendritic plasticity support running-improved spatial learning and depression-like behaviour in stressed rats. *PLoS One, 6*, 9. https://doi.org/10.1371/journal.pone.0024263.

Ziegelstein, R. C. (2017). Personomics: The missing link in the evolution from precision medicine to personalized medicine. *Journal of Personalized Medicine, 7*(4), 11. https://doi.org/10.3390/jpm7040011.

Chapter 4
The Benefits of Non-pharmacological Interventions

1 Introduction

Individual use and positive feedback about a treatment do not guarantee its effectiveness and safety. Such evidence must be provided by rigorous studies before the treatment can be recommended and correctly implemented. The number of intervention studies on non-pharmacological interventions (NPIs) has exponentially increased over the last decade. This chapter presents some of these studies, positive or not, on targeted health indicators and on the five categories of NPIs. Of course, the format of the book precludes a complete inventory and exhaustive comparison between NPIs. Examples will be given to illustrate the prevailing logic and the trends that have been emerging since 2010.

2 A Wind of Optimism

More than 50 years of effort and colossal financial investment in clinical research by academic authorities and private companies have enabled the development of a specific consensual framework dedicated to drug validation and surveillance. The standardized guide of the 11,000 drugs available on the market is adapted for users and professionals and is updated frequently. In the drug field, declaration processes are established, manufacturing standards are fixed, indications are standardized, dosages are formalized, doctors must alert authors in case of incident/accident, and authorities have to monitor and surveil their use.

Nothing like this exists to date in the NPI field. Current progress is dedicated to the collection and analysis of data coming from intervention studies published in the best scientific and medical journals.

2.1   Interest of Researchers and Academic Institutes

An exponential increase in the number of clinical studies and systematic reviews analyzing the benefits and risks of NPIs for certain pathologies is evident in the medical and scientific literatures. For example, in Alzheimer's disease, NPIs are recommended by many national and international health organizations in the fight against Alzheimer's disease, particularly in the early stages of the disease (California Workgroup 2018; World Alzheimer Report 2018; French High Authority for Health 2011, 2018; UsAgainstAlzheimer's 2019). Unfortunately, many people in the world suffering from this disease do not benefit from NPIs enough; no doubt because the decision-makers are waiting for a miraculous treatment that does not materialize.

> **One New Case of Dementia Every Three Seconds Around the World: Time for NPIs**
> The Professor of Social Policy at the London School of Economics, Martin Knapp, mentioned in the World Alzheimer Report 2018: *"We're better at recognizing and assessing people's needs and hopefully at least helping them with a label for the distress they're experiencing. And there are some interventions that are being developed in a psychosocial area, around, for example, cognitive stimulation therapy and cognitive remediation and those seem, on average, to work for people in the mild to moderate stage of dementia, in terms of just preserving that cognitive ability for a bit longer"* (World Alzheimer Report 2018, p. 27).

NPIs have proven themselves to delay the onset of the disease, to treat its symptoms, to increase patient autonomy, to improve quality of life, and to relieve caregivers. Some NPIs are more effective, less risky, more relevant, more acceptable, and more profitable than others as some systemic reviews are starting to demonstrate, for example, in relation to Alzheimer's (Cammisuli et al. 2016).

The number of studies is so significant nowadays that NPIs have to move away from simple impersonal hygienic-dietetic advice that mixes uses, practices, pointless solutions, conflicts of interest, sectarian aberrations, and fake medicines. International reports urge medical and scientific societies to rely on recent studies and break the shackles of restricting the notion of the efficacy of a health solution to the double-blind randomized controlled trial, and the benefit to total healing or increase of survival. As an example, the French National Authority for Health explains in an official document dating from 2011 that *"several NPIs are possible. They are, both on an outpatient basis and in an institution, an important element of therapeutic care. However, due to methodological difficulties, none of these interventions has demonstrated its efficacy"* (HAS 2011, p.19). There are countless meta-analyses and systematic reviews in the international scientific and medical literatures on the medical relevance of NPIs for people suffering from Alzheimer's disease, at risk of Alzheimer's disease (for example, mild cognitive impairment), or related dementia. If clinical studies on proposed NPIs alone or in combination are still lacking (Karssemeijer et al. 2017), the data currently available are sufficient to conclude that the benefit-risk and cost-effectiveness of some NPIs are better than others (Cohen-Mansfield 2013; Anderson et al. 2017; Nickel et al. 2018). An adapted physical activity program, a therapeutic education program, behavioral intervention, and cognitive intervention are essential. NPIs like aromatherapy deserve more study. On the other hand, manual therapies, nutritional therapies (diet or food supplement), or video games have not proven to be effective despite a significant buzz on the Internet (UsAgainstAlzheimer's 2019). Reviews such as those by Burckhardt et al. (2016) on the role of omega 3 and also that of Ballard et al. (2011) are clear on this subject.

Even systematic reviews exist today. That of Horneber and collaborators (2012) published in the *Integrative Cancer Therapies* journal relates to NPIs relevant in the case of cancer.

2.2   Methodological Advances for Studies

A strong positive signal is the clear improvement in the methodological quality of the studies, their quantity, and the number of countries involved in the last 20 years. Previously, studies were essentially observational and mechanistic. They have more recently become interventional and follow international standards: upstream protocol declaration; validation by an independent ethics committee; relevant choice of outcome; measurement using reliable, sensitive, and valid instruments; adapted statistical or qualitative analysis; more numerous participants; and homogeneous groups and better described NPIs (Chap. 8).

Meta-analyses include more and more recent intervention studies, positive or negative. They point out the methodological limits (e.g., weak design, participants' heterogeneity, ethical conditions, statistical errors, speculative interpretations, conflicts of interest, publication bias due to predominance of English), encouraging researchers and promotors to improve the quality of studies.

2.3   New Outcomes

An additional strong signal is that studies no longer focus solely on subjective markers such as health-related quality of life, pain, anxiety, depression, quality of sleep, fatigue, dyspnea, or self-esteem. They incorporate objective markers such as biological markers (e.g., blood measures, inflammation rate), epidemiological markers (e.g., disease-free survival, survival without recurrence of cancer), cognitive markers (e.g., performance in executive function tests), behavioral markers (e.g., number of steps for a week), and, finally, economic markers (e.g., number of hospitalizations, medical visits). This intersection of indicators gives more consistent results to enable the recommendation or otherwise of an NPI.

3 Targeted Health Goals with NPIs

3.1 Anxiety

Meta-analyses have looked at the effectiveness of a specific NPI in treating anxiety during breast cancer treatment. The meta-analysis of Haller et al. (2017) assesses mindfulness-based stress reduction (MBSR) in addition to biological treatments for breast cancer on the basis of 10 studies that included 1709 patients. Compared to routine care, benefits of the MBSR program are positive for anxiety. They last between 6 months and a year after the treatments. Why deprive patients of this?

3.2 Back Pain

Back pain affects about 10% of the population in each nation. This persistent pain is called "mild or nonspecific low back pain" and causes movement limitations, increased physical inactivity, recurring negative thoughts, emotional difficulties, and repeated work stoppages. People can feel so helpless that almost 60% do not think they can recover. They multiply treatments and care strategies without lasting benefit. Surgeries can be performed, sometimes without much success over time. The costs of care and days off from work are soaring, with significant costs borne by the patients themselves. Back pain was responsible for the loss of 150 million working days in England in 1993.

NPIs will target different areas:

- Muscular (such as strengthening the postural muscles)
- Tendon and joint (such as improving flexibility)
- Neurological (such as pain regulation)
- Postural (such as ergonomic adjustment of the workstation)
- Psychological (such as pain desensitization and stress regulation)
- Social (such as relationship and work organization)
- Environmental (such as bed replacement)

3.3   Specific Method of Osteopathy

A British randomized controlled trial has evaluated the benefits of an osteopathic method dedicated to treat patients with acute and subacute spinal pain (Williams et al. 2003). The study included 201 persons between 16 and 65 years of age who had consulted their general practitioner for neck or back pain that had appeared between 2 weeks and 3 months previously. All of the patients in the trial continued to receive their usual treatment such as advice on rest, daily activity, and work, and if necessary a prescription for non-steroidal analgesics and anti-inflammatory drugs, and an orientation toward secondary treatments (e.g., physiotherapy). The assessments between the osteopathy and control group were carried out before the start of the intervention, at the end (+2 months) and again at 6 months. The participants followed 3–4 sessions carried out by a doctor trained in the NPI. The sessions were spaced 1–2 weeks apart with a maximum time interval of 2 months. The results at the end of the 2 months of osteopathic intervention showed a decrease in the intensity of spinal pain and an improvement in the mental component of quality of life compared to the control group. The mental component of quality of life remained superior to the control group after 6 months. In contrast, pain levels became equivalent between the groups after 6 months.

3.4   Specific Method of Psychotherapy

A randomized controlled trial in Sweden evaluated the effectiveness of a primary prevention NPI conducted with workers and their employers to prevent disabilities due to a musculoskeletal disorder (Linton et al. 2015). The program was compared to usual care. One hundred and forty people aged 27–65 years participated in this study. They suffered from low back pain with a high risk of developing chronic musculoskeletal disorder. The researchers assessed days off work related to musculoskeletal conditions, use of care, perceived health, and the intensity of pain. The measurements were taken before and after the intervention, then 6 months after. The patients in the experimental group received three sessions of cognitive behavioral therapy by a trained psychologist. Each session lasted between 60 and 90 min. The study shows a halving of the average number of days off work 6 months after the intervention (17 days instead of 38 days). This difference was statistically significant compared to the control group at 6 months. The researchers also noted an improvement in perceived health and a decrease in the use of care for the intervention group compared to the control group. The intensity of the pain felt was reduced identically in the two groups.

3.5   Specific Method of Yoga

A meta-analysis showed the benefits of a specific form of yoga on low back pain (Wieland et al. 2017). Yoga includes physical exercises, breathing exercises, relaxation techniques, and meditation. Yoga increases mindfulness and mind-body connections.

3.6   Recommendations

Many other clinical trials showed the effectiveness of specialized NPIs in the treatment and prevention of chronic low back pain such as Pilates, Dru yoga, Back School method, and McKenzie method. These isolated studies compiled in systematic reviews made by independent researchers (e.g., Wieland et al. 2017) and academic colleges (e.g., Chou et al. 2017; Qaseem et al. 2017) support the dissemination of relevant NPIs.

Recommendations of the American College of Physicians on Treatment of Adults with Acute, Subacute, or Chronic Low Back Pain (Qaseem et al. 2017)

A committee of the American College of Physicians explored systematic reviews published through November 2016 including reduction or elimination of low back pain, improvement in back-specific and overall function, improvement in health-related quality of life, reduction in work disability and return to work, global improvement, number of back pain episodes or time between episodes, patient satisfaction, and adverse effects.

Recommendation 1: *"Given that most patients with acute or subacute low back pain improve over time regardless of treatment, clinicians and patients should select nonpharmacologic treatment with superficial heat (moderate-quality evidence), massage, acupuncture, or spinal manipulation (low-quality evidence). If pharmacologic treatment is desired, clinicians and patients should select nonsteroidal anti-inflammatory drugs or skeletal muscle relaxants (moderate-quality evidence) (Grade: strong recommendation)."*

Recommendation 2: *"For patients with chronic low back pain, clinicians and patients should initially select nonpharmacologic treatment with exercise, multidisciplinary rehabilitation, acupuncture, mindfulness-based stress reduction (moderate-quality evidence), tai chi, yoga, motor control exercise, progressive relaxation, electromyography biofeedback, low-level laser therapy, operant therapy, cognitive behavioral therapy, or spinal manipulation (low-quality evidence) (Grade: strong recommendation)."*

3.7 Depression

Researchers have compared several NPIs to treat depression in general medicine. A meta-analysis based on 11 trials including 1041 patients has compared cognitive behavioral therapy (CBT), physical activity program, problem-solving psychotherapy, behavior change program, and light therapy (Holvast et al. 2017). The authors conclude that CBT is preferable for treatment of depression compared to other NPIs while encouraging further studies to confirm the findings.

CBT is the most commonly researched psychological therapy for the treatment of people with persistent subthreshold depressive symptoms or mild to moderate depression. A meta-analysis of 75 trials comparing CBT and control groups showed significant differences in a multivariate meta-regression analysis. No indication was found that CBT was more or less effective than pharmacotherapy. Combined treatment was significantly more effective than pharmacotherapy alone. The authors concluded "there is no doubt that CBT is an effective treatment for adult depression" (Cuijpers et al. 2013). Evidence suggests that CBT is an efficacious treatment for depression, and its use is recommended by the National Institute for Health and Care Excellence (NICE 2009).

CBT: A Depression Therapy for All Ages?
A meta-analysis of 366 randomized clinical trials including 36,072 patients comparing psychotherapy with control conditions found that psychotherapies had lower effect sizes in children and adolescents compared with adults, and no significant differences were found between middle-aged and older adults (Cuijpers et al. 2020).

3.7.1 Cognitive Functions in Persons with Dementia

Yorozuya et al. (2019) reviewed NPIs in improving the global and specific cognitive function in older people with dementia living in nursing homes. The study aimed to investigate the effects, contents, frequency, duration, length, and form of NPIs to determine the most effective interventions. The systematic search of peer-reviewed literature published between January 2008 and October 2018 performed on PubMed, Cochrane Library, Web of Science, and CINAHL databases indicated six randomized and one nonrandomized controlled trials. Authors concluded that NPIs improve global cognitive function, executive function, attention, memory, and constructional apraxia in older people with dementia living in nursing homes. They also concluded that global cognitive function and memory benefits are highly reliable based on high-quality trials. The combination of exercise, cognitive training, and activities of daily living, involving an intervention of at least 30 min at least three times a week over at least 8 weeks using the integrated form, is recommended for improving global and specific cognitive functions. Authors recommended preferential intervention dedicated to specific cognitive functions such as memory, executive function, visuospatial function, and constructional ability in persons with dementia.

3.8 Fatigue

A number of studies have focused on reducing fatigue during cancer treatments. A meta-analysis, which selected 17 randomized controlled trials including 1380 patients, showed that an NPI offering moderate-intensity physical activity of 2 h per week during breast cancer treatments significantly reduces fatigue (Carayol et al. 2013). Inverse dose-response relationships were observed for fatigue and quality of life (QoL), supporting the hypothesis of greater improvements in fatigue and QoL with lower weekly prescribed exercise doses (< 12 Metabolic Equivalent of Task h/ week). The findings underlined a program targeting 8–10 Metabolic Equivalent of Task hours per week for patients with breast cancer receiving adjuvant therapy.

An effective and safe exercise program during breast cancer could consist of one resistance session for principal muscular groups and two moderate-intensity 30–45-min aerobic sessions per week (Carayol et al. 2019). The program includes exercise sessions planned thrice weekly including individually supervised hospital-based exercise sessions and non-supervised home-based sessions. Hospital-based supervised exercise sessions are planned every 3 weeks on the same days as chemotherapy and radiotherapy. The exercise program is delivered by a specialist. Exercising while receiving chemotherapy or radiotherapy involves specific brakes. In particular, the loss of physical condition induced by chemotherapy in patients with breast cancer is an important brake on performing relatively high doses of exercise as compared to exercising after adjuvant therapy as a cancer survivor. The patients participate in one muscle strength session and two aerobic sessions each week. Every session starts with a warm-up for 10 min including joint rotations (fingers, wrists, elbows, shoulders, neck, hips, knees, ankles, feet, and toes) for 3 min, slow jogging for min, and cross-over stepping, thighs lift, and heels to buttocks for 2 min. Every session ends with a cool-down followed by flexibility exercises (10 min). Strength sessions (once a week) target six main muscle groups (hamstrings, quadriceps, buttocks, abdominal, back, shoulders/arms) by asking patients to achieve six different tasks. Each task is performed for 2–5 sets with 6–12 repetitions. Two to five different tasks with increasing difficulty are available for each muscle group. Every 6 weeks the exercise specialist proposes a two-repetition or one-set increase, and/or a shift to a more difficult task according to the patient's physical condition and progression. Aerobic exercise sessions are performed at moderate intensity and adapted to the patient's physical condition and progression in the range of 50%–75% of the maximum heart rate for 30–45 min. Initial exercise intensity is individualized but generally begins at 50%–55% of the maximum heart rate and progress to 65%–75% of the maximum heart rate by weeks 20–26. Initial exercise duration is also individualized but generally begins at between 25 and 35 min per session, achieving 40–50 min per session by weeks 20–26. The starting point for exercise prescription is determined according to the maximum heart rate estimated as 220 – age of the patient, and patient's exercise history and current practice. Rate of progression is individualized according to the severity of treatment-related side effects. When patients present health troubles, important fatigue, or any symptom that could limit exercise, an adapted decreased dose is proposed to patients by the exercise specialist. Supervised hospital-based sessions are achieved on a cycloergometer. For home-based practice, various modalities of aerobic exercise (e.g., walking, jogging, cycling, dancing/fitness, swimming) aid adherence to the program and generate pleasure. Hospital-based supervised exercise sessions aim to provide patients with relevant instructions to allow reproducibility at home and an increase in autonomy. Every supervised session is based on theoretically grounded specific behavioral targets (e.g., problem-solving barriers, self-efficacy, social support) and behavior change techniques (e.g., providing information about behavior-health links and the consequences of practice, providing instruction and demonstrations of the behavior, setting graded tasks, prompting self-monitoring of behavior, and planning social support or social change) to improve behavioral change and the patient's adherence. Hospital-based supervised exercise sessions are scheduled on the same day as chemotherapy administration and during radiotherapy, at the frequency of one every 3 weeks. In total, nine hospital-based supervised exercise sessions are planned in the course of the intervention. Each session lasts approximately 50–70 min. Home-based non-supervised sessions are planned three times per week, except that only two home-based sessions a week are scheduled on the weeks involving one supervised hospital-based exercise session. Precise written instructions for home-based sessions are given to patients in a personal educational workbook including information about disease and reasons for being physically active during active treatment for cancer, written instructions illustrated with pictures for performing home exercises (e.g., warm-up content, muscle tasks, prescribed number of series/repetitions with space for potential incrementation, aerobic intensity and duration, flexibility tasks), schedule for planned home-based sessions, and a patient log to evaluate patient adherence. The strength-based exercise that is taught to participants during the supervised sessions does not need any particular material so that it can easily be done at home. Patients are asked to fill in the adherence log at home, recording whether planned sessions were achieved or not, number of achieved muscular exercises, duration of session, rating of perceived exertion (on a scale of 1–10), reason for missed session, and any commentary that they would like to discuss with the exercise specialist at the next supervised session. The workbook includes a range of behavior change techniques (e.g., providing information about behavior-health links and the consequences of practice, providing instruction and demonstrating the behavior, setting graded tasks, and prompting self-monitoring of behavior).

3.9   Fibromyalgia

Researchers have suggested there are therapeutic benefits of tai chi programs in patients with fibromyalgia. A Chinese randomized controlled trial has compared a Yang-style tai chi program to control fibromyalgia with an intervention including wellness education and stretching (Wang et al. 2010). The tai chi intervention involving a 60-min session supervised by tai chi master with more than 20 years of experience took place twice a week for 3 months. The first session presents the theory behind tai chi and its procedures and provides participants with printed materials on its principles and techniques. The following sessions invite participants to practice 10 forms from the classic Yang style of tai chi. Each session includes a warm-up and self-massage, followed by a review of principles, movements, breathing techniques, and relaxation. Participants are instructed to practice tai chi at home for at least 20 min each day. At the end of the 3-month intervention, participants are encouraged to maintain their tai chi practice, using an instructional DVD, up until the follow-up visit after 6 months. The primary end point of this study involved a questionnaire to establish the impact score for change in the patient's fibromyalgia at the end of 3 months. Secondary end points included scores from a health-related quality of life questionnaire. The assessments were repeated at 6 months to test the durability of the response. The results showed that 33 of the 66 randomly assigned patients in the tai chi group had clinical improvements in fibromyalgia impact and health-related quality of life score. No adverse events were observed. Tai chi may be a useful treatment for fibromyalgia.

3.10   Knee Arthritis

An English study by Woods et al. (2017) focuses on NPIs effective against osteoarthritis of the knee based on 88 randomized controlled trials that included 7507 patients. The study compares different NPIs: acupuncture, orthopedic devices, heat treatment, insoles, interference therapy, light therapy, manual therapy, neuromuscular electrical stimulation, pulsed electrical stimulation, pulsed electromagnetic fields, static magnets, and transcutaneous electrical nerve stimulation. The authors conclude that acupuncture is the NPI with the best cost-benefit ratio on quality of life (Woods et al. 2017).

3.11   Obesity

Overweight is a major public health problem. In the United States, more than 60% of the population is overweight; 34% is obese. Being overweight increases the risk of developing a chronic disease or a cancer. Dietary supplements based on soy lecithin and green tea are natural foods. They focus on several mechanisms such as improving satiety, thermogenesis, and inhibition of the two enzymes metabolizing carbohydrates and lipids. A randomized controlled trial in the United States tested the effectiveness of a dietary supplement based on soya lecithin and green tea on low-calorie diet and weight loss. The diet asked overweight and obese participants to reduce their daily food intake by 30% or

500 kcal compared to their usual consumption. The trial included 50 healthy people, aged 18–59 years, whose body mass index was between 25 and 40 kg/m². Participants were randomly assigned either to the intervention group receiving the supplement or to the control group receiving a placebo. The food supplement contained a mixture of 40 mg of natural soy phospholipid, 35 mg of green tea extract, and 25 mg of mixed phospholipids. Participants took 300 mg a day, one tablet taken 1 h before breakfast and two tablets taken 1 h before evening meals for 8 weeks. The study showed that, during the first month, the daily intake of food supplements improved compliance with the low-calorie diet, mood, and subjective fatigue compared to the placebo group. This benefit was no longer seen from the second month. No change in body weight or fat mass was obtained.

3.12 Pain

A review by Volkow and McLellan (2016) focused on opioid abuse for chronic pain. The authors recommend non-pharmacologic treatments such as cognitive behavioral therapy (CBT), exercise therapy, and complementary medicine (e.g., yoga, meditation, acupuncture).

3.13 Post-traumatic Stress

Exposure to traumatic experience is a common health condition: 60% of men and 51% of women in the general population have experienced at least one traumatic episode in their lives. For men, experiences are related to war or physical attack. Among women, physical aggression by their partner is the most common cause. An American randomized controlled trial by Van der Kolk and collaborators, published in the *Journal of Clinical Psychiatry* in 2007, compared the effectiveness of the eye movement desensitization and reprocessing (EMDR) therapy to an antidepressant treatment (fluoxetine) and a placebo for 6 months to reduce the severity of the symptoms of post-traumatic stress disorder.

Defining Eye Movement Desensitization and Reprocessing (EMDR)
EMDR is a psychotherapeutic method in which patients make eye jerks while thinking of a traumatic experience. This NPI allows patients to follow their own path, to move freely, to be sensitive to their internal and cognitive sensations.

The study looked at 88 adults diagnosed with post-traumatic stress disorder. Patients were randomly assigned to either the EMDR group, the fluoxetine group, or the placebo group. The evaluations were carried out at the beginning and at the end of the intervention at 8 weeks, then after 6 months at follow-up. EMDR specifications were six individuals maximum, 90-min sessions, and specific training for psychotherapists. The effectiveness of NPI was greater than that of the placebo in reducing symptoms of post-traumatic stress disorder. The EMDR program was as effective as the antidepressant in reducing symptoms between the beginning and the end of the EMDR protocol (Van der Kolk et al. 2007). The NPI was more effective than the antidepressant in reducing symptoms of post-traumatic stress disorder and symptoms of depression in the long term, 6 months after treatment. In addition, at 6 months, 58% of patients in the EMDR group reached an asymptomatic state compared to 0% in the antidepressant group.

4   Psychological Interventions

Art therapy programs like music therapy treat behavioral problems in elderly people. A systematic review assessed the effects of an NPI of music therapy in people with dementia at the end of treatment and at least 1 month after the intervention (Van der Steen et al. 2018). It included 22 randomized controlled trials with 890 participants, the control groups having either received usual care or activities with or without music. The authors concluded that the 5-session music therapy program had a significant effect, with the reduction of depressive symptoms and the improvement of behavioral disorders at the end of treatment. The authors remain more skeptical about the benefits in terms of agitation, aggression, and cognition.

Specialized hypnotherapy, as a psychological intervention, is one of the most effective in smoking cessation. A hypnotherapy program is also an efficient solution for anesthesia for breast cancer surgery (Amroui et al. 2018).

Hypnotherapy Program for Smoking Cessation
A 3- to 4-session hypnotherapy program provided by an NPI-trained health professional allows some smokers, even the most reluctant, to quit smoking.

A hypnosis session has three phases. The induction phase focuses the patient's attention on a point or on the therapist's voice in order to create maximum relaxation and a state of confusion. It allows the patient to enter a modified state of consciousness, which awakens his senses, mobilizes his internal resources, softens his certainties, which then opens him up to suggestibility, making him indifferent to the outside world. In this state, the patient shows all of the characteristics of a profound relaxation, with deep breathing, an emotionless expression, a slightly confused voice, and speaking rarely. The work phase changes his feelings, his perceptions, and his gestures with regard to tobacco through suggestions and exercises that can combine images. By sharpening critical thinking and sensory perception, the hypnotherapist opens the patient to new sensations and a new way of understanding tobacco that is less glamorous, less idealized, less a vector of pleasure, and more like a poison. The last phase brings the patient back to the normal waking state.

Table 4.1 The Otago program to prevent falls in persons over the age of 70

Health condition	450,000 accidental falls each year in people over 65 in France. First cause of death by accident (9412 in 2008). The fall causes injuries, trauma, sequelae, loss of capacity, pain, loss of self-confidence, loss of autonomy. At 80, walking is no longer enough to prevent falls.
NPI	6-month adapted physical activity program, with 3 sessions of 30 min per week. Posture and balance exercises (on one or two feet, forward or backward, standing or not standing), muscle strengthening (static or dynamic, sitting or standing, with body weight or an additional load on the ankles), walking flat (with an obstacle, heel-toe or normal amplitude of the steps, legs spread or aligned, straight or with rotation, forward or backward).
Mechanisms	Reconditioning of the balancing function (improvement of the perceptual and motor sub-functions) and of the muscle function (muscle gain, improvement of proprioception).
Benefits	47% reduction in the risk of falling 1 year after the initial program. 35% decrease in the number of falls in the year after the program. 35% decrease in the number of injuries in the year after the program. 70% of participants continue to practice 1 year after the end of the program.
Risks	Rare wounds, bruises and fractures.
Best practices	Expected progression according to 4 levels of mastery. Pain-free practice for osteoarthritis, bone-articulation and muscular-tendon disorders, osteoporosis, or painful disease. Monitoring for chest pain, breathing difficulties, dizziness, persistent muscle pain, and falls. Use a hip pad in the event of repeated falls.
Professional	Adapted physical activity teacher or trained physiotherapist.
References	Campbell et al. (1997), Robertson et al. (2001), Campbell and Robertson (2007)

Note: This table, like the following ones, summarizes the current state of knowledge on an NPI aiming to solve a targeted health problem on the basis of the best published studies. It is by no means a recommendation for individual use. Only a doctor is authorized to do so

Therapeutic educational programs, also called disease management protocols, facilitate adherence to long-term drug therapies (WHO 2003). Ergonomic pillows prevent sleep disorders, hypoxic masks prevent physical deconditioning, and socks facilitate venous return. Horse therapy improves the social functioning of people with autism.

5 Physical Interventions

Adapted physical activity (APA) programs cure mild type 2 diabetes. Others prevent falls in people over the age of 70 such as the Otago program (Table 4.1).

Hortitherapy programs improve cognitive parameters. Manual therapies treat back pain (Rakel 2018). Spa treatments treat gonarthrosis. Qi Gong, according to a systematic review, reduces resting systolic blood pressure by approximately

18 mmHg and diastolic blood pressure by approximately 9 mmHg in patients with hypertension (Lee et al. 2007).

Auriculotherapy

Auriculotherapy is an NPI that consists of stimulating specific areas of the ears, either by pricking or by another type of stimulation (electrostimulation, massage, magnetic field, laser, etc.) Exploratory works indicate clinical interest of this NPI to treat dependence on tobacco or alcohol, pain, dermatological problems, nausea, preoperative anxiety, sleep disorders, allergic rhinitis, post-cesarean hypo-galactorrhea, and endometriosis. A French analysis based on 43 randomized controlled trials and a Cochrane meta-analysis have been conducted (Gueguen et al. 2013). The total number of participants in each group had to be more than 25, regardless of age. The patients studied had to have a pathology diagnosed by a doctor. Trials combining auriculotherapy with other NPIs or other treatments were not included. The auriculotherapy intervention was compared to a control group without treatment, placebo, or using another active treatment. Techniques used included stimulation with needles, semi-permanent needles and semi-permanent pins, micro-bleeding, electro-acupuncture, laser, magnetic stimulation, and massage. The number of sessions varied from a single session to 40 sessions spread over a maximum of 8 weeks depending on the study. The authors concluded that auriculotherapy is effective in pre-operative pain and anxiety (Gueguen et al. 2013). Studies of other health conditions are too heterogeneous and have too many methodological limitations to conclude that there is a benefit. Auriculotherapy reveals minor and transient undesirable effects such as vagal discomfort (quite frequent), dizziness (rare), and chondrites (exceptional).

6   Nutritional Interventions

Food supplements like vitamin D, omega 3, and creatine prevent the risks of osteoporosis, cardiovascular disease, and muscle wasting, respectively, with advancing age. Diets like the Dash diet lower the risk of a heart attack by 37%. The gluten-free diet treats people with celiac disease (Al-Toma et al. 2019) (Table 4.2).

A clinical trial, initiated in 2012 and published in the journal *Lancet Neurology*, tested the long-term effectiveness of the consumption of a dietary supplement based on ginkgo biloba on reducing the risk of the onset of Alzheimer's in older adults who complain about their memory (Vellas et al. 2012). The randomized controlled trial evaluated daily consumption of ginkgo biloba extract in people over the age of 70 who complain about their memory. These complaints are often associated with an increased risk of Alzheimer's disease. One possible mechanism of action was the antioxidant effect of this natural extract. The hypothesis was that elderly people with memory difficulties consuming ginkgo biloba extract could delay the onset of

Table 4.2   Gluten-free diet for celiac disease

Health condition	Celiac disease is a genetically determined chronic inflammatory enteropathy triggered by gluten. It is one of the most widespread food intolerances in the world. It affects approximately 1% of the population. Women are twice as affected as men. The diagnosis is often late, between 40 and 45 years of age. The disease manifests as diarrhea, weight loss, fatigue, bloating, stomach ache, nausea and, in children, stunted growth.
NPI	Gluten-free diet, for life, under medical prescription.
Mechanisms	In genetically predisposed patients, the absorption of food products containing gluten (found in many cereals, such as wheat, barley, spelt, rye, kamut, and triticale), even if they are only traces, causes an immune reaction of the small intestine. This results in chronic inflammation of the mucous membranes of the small intestine and atrophy of the epithelial cells. These characteristic histological changes can induce a malabsorption syndrome. The gluten-free diet regenerates the mucous membranes of the small intestine.
Benefits	Disappearance of symptoms, prevention of complications.
Risks	Deviation from the diet which can lead to complications.
Best practice	Regularly perform medical examinations to update and rule out any deficiencies and complications.
Professional	Dietitian following medical diagnosis

the disease by a few years. The study included 2820 people over the age of 70 whose memory complaints were verified during a medical examination. The follow-up lasted 5 years. Participants were randomly assigned to either the group receiving ginkgo biloba extract (120 mg of ginkgo biloba extract in tablet form, twice a day, for 5 years; the tablets were provided during doctor visits every 3 months) or the placebo group receiving a tablet of identical appearance, smell and taste. Cognitive, functional, and depressive states were assessed each year. Safety was assessed every 3 months by monitoring vital signs and performing physical and neurological examinations. Possible side effects have also been reported. The study showed that after 5 years, 70 participants out of 1406 in the group receiving ginkgo biloba extract were diagnosed with Alzheimer's disease compared to 84 out of 1414 in the placebo group. This difference of 14 people is not statistically significant. Daily consumption of 240 mg of ginkgo biloba extract for 5 years does not protect people over 70 years of age from Alzheimer's disease.

7   Digital Interventions

Digital methods, including an electronic cigarette, can result in cessation of smoking. Serious games facilitate the rehabilitation of accident victims. Twelve to 13% of the world's population is affected by a specific phobia causing excessive anxiety. Virtual reality systems have been shown to successfully treat phobias that occur in elevators (Rahani et al. 2018) as mentioned in Table 4.3.

Table 4.3 Phobia therapy with virtual reality

Health condition	When phobia becomes disabling on a daily basis, it can be the subject of a protocol in order to cure it; for example, the fear of spiders.
NPI	A virtual reality headset is offered in addition to cognitive behavioral therapy. The therapy lasts 6–8 weeks. Equipped with a helmet, glasses, and a motion sensor, the patient is gradually exposed to the object of his phobia. His entire field of vision is immersed in a virtual world which has no connection with reality. He becomes an actor of another three-dimensional world, which changes according to his head and body movements.
Mechanisms	The entire field of vision is blocked by stimuli; a progressive desensitization of the phobic object occurs.
Benefits	Virtual reality is effective in 80% of cases. While this is less effective than the 90% from cognitive behavioral therapies alone, the treatment time is shorter and the benefits persist longer.
Risks	10% of patients complain of discomfort. They experience discomfort during or after immersion in virtual reality. This usually occurs in people who are sensitive to motion sickness. There is a gap between the body's balance system and that of vision.
Best practices	If virtual reality hardware has become accessible, software remains expensive.
Professional	Trained psychotherapist.

8   Elemental Interventions

Phytotherapeutic products are used to treat mild depressive disorders, and essential oils play a role in reducing the effects of stress. Herbal medicine, or herbal therapy, has been practiced in various forms for thousands of years and remains the only medicine used in certain places of the world. St. John's Wort is an example (Apaydin et al. 2016) (Table 4.4).

Cosmeceuticals treat dysmorphophobias, creams provide protection from the sun, and wigs help patients during chemotherapy for cancer. Ultrasound treats tendonitis, and light therapy programs reduce symptoms of seasonal depression. Lithotherapy methods soothe neck pain.

9   An NPI Guideline Still Under Construction

Three well-known books on the subject of NPIs internationally, written by Edzard Ernst et al. (2008), Marc Micozzi (2019) and David Rakel (2018), indicate the benefits of biofeedback for high blood pressure; St. John's Wort for depression; fish oils rich in acids, omega 3, and fat for hyperlipidemia; and valerian for insomnia. But the authors remain cautious and regret that no international classification exists to date.

Table 4.4 St. John's Wort for mild to moderate depression

Health condition	Depression affects 3 million people in France, affecting twice as many women as men. It is a debilitating disease that causes great suffering that can lead to suicide.
NPI	St. John's Wort capsule (standardized extract of 3% hyperforin or 0.2% to 0.3% hypericin used in clinical trials). The prescription is to take 300 mg, 3 times a day for solid extracts, with meals.
Mechanisms	Acts on neurotransmitters (especially serotonin).
Benefits	Compared to taking placebo, St. John's Wort is more effective in improving the depressive symptoms of mild to moderate depression. The effect manifests after 1 month. Compared to taking conventional tricyclic antidepressants (selective serotonin reuptake inhibitor), St. John's Wort is equally effective and is associated with a lower risk of stopping treatment due to side effects (Linde et al. 2008).
Risks	St. John's Wort stimulates enzymes in the liver, which can degrade many drugs before they have time to work. It is not recommended in conjunction with antiretroviral, anticoagulant, or cancer treatment, or with contraceptive use.
Best practices	It is important to be attentive to the concentrations of the active ingredient in the various preparations of St. John's Wort available on the market. Depression is a disease that must be diagnosed according to specific criteria by a doctor. Whatever its intensity, depression requires medical attention. Self-medication is not recommended. It is recommended to gradually decrease the doses of St. John's Wort for 1–2 weeks to interrupt treatment to avoid the risk of withdrawal syndrome. It is advised to not exceed the recommended daily dose, and to keep out of reach of children.
Professional	General practitioner or psychiatrist.

The fifth version of Marc Micozzi's book is 736 pages with 44 chapters, where the classification of NPIs is mainly by continent (Micozzi 2019). The first version of Edzard Ernst's book is 424 pages long with 7 chapters, including a separation between alternative therapies and complementary medicines (Ernst et al. 2008). The fourth version of David Rakel's book is 1122 pages with 118 chapters and without full references, where many NPIs are components of an integrative therapy, rarely with any particular order of relevance (Rakel 2018). How do you fit about 10,000 NPIs in a single book without making shortcuts? How is it possible to keep up-to-date with advances in this science? Marc Micozzi strives to update his book every 5 years. How is it possible to achieve this alone?

For the moment, no synthesis is available that is carried out with the greatest scientific rigor on the indications, the modes of use, and exhaustive comparisons. Various attempts mostly comprise opinions with, in some cases, evidence of conflicts of interest. For example, an attempt is made by the French National Olympic and Sports Committee to establish recommendations for sport therapy programs. It is astonishing to find a priority among the five recommended practices for breast cancer, sports like rafting or canoeing, which are not very accessible, and if practiced frequently can be potentially traumatic following surgery.

Caution Regarding Systematic Reviews

Most meta-analyses, systematic reviews, and collective expertise agree that one must remain cautious with regard to the comparative recommendation of NPIs because:

- An extensive list of NPIs does not exist to date
- The full description of NPIs is uneven in the available studies
- The methodologies for evaluating NPIs are heterogeneous
- Studies in prevention and curative treatment are less numerous than those in complementary care of biomedical treatments
- Studies available are mainly North American (not only because scientific journals and the main medical databases are in English, but also because research funding is higher there than in any other continent). With the resulting extrapolation of findings, and NPI implementation in other continents, it is important to keep in mind the American culture and way of life of the participants included in studies.

At the same time, the transparent and rigorous academic procedures are comprehensive and fastidious, such as the National Institute for Health and Care Excellence (NICE) commissioning the National Collaborating Centre for Mental Health to update the clinical guidelines on depression in adults that published in October 2009 (Table 4.5).

9.1 The More NPIs Are Implemented, the More Systematic Review Conclusions Are Prudent

Intervention studies on NPIs are very different from each other, especially in which humans are conducting the intervention. There is too much variability in studies, both in those focusing on prevention and therapy, to be able to compare strategies for identifying NPIs, their objectives, and effects. In addition, the designs generally have methodological flaws, limiting them from accurately determining which interventions could be applied in prevention or clinical practice. For example, Riquelme-Miralles et al. (2019) examined NPIs to improve therapeutic adherence in the case of tuberculosis for clinical practice. A systematic review in Medline and Embase was performed. Thirty-seven articles were analyzed. The authors concluded that the NPIs were disparate, grouped into education, psychological interventions, new technologies, directly observed treatment, incentives, and improved access to health services. In the treatment of latent infection, the majority of studies were conducted

Table 4.5 National Collaborating Centre for Mental Health's work in progress to update the clinical guideline on depression in adults published by NICE in October 2009 (NICE 2009)

Date	Update
13 May 2020	Stakeholder list updated
26 February 2020	Review documents
03 December 2019	Declaration of interests
31 October 2018	Scope published
31 October 2018	Consultation comments published
15 May 2018–12 June 2018	Addendum consultation
31 October 2017	Committee list updated
18 July 2017–12 September 2017	Draft guidance consultation
12 July 2017	Pre-consultation documents released
04 March 2016	Equality impact assessment published
09 June 2015	Committee meeting
19 June 2015	Call for evidence on specific clinical questions
31 March 2015	Consultation on the draft scope
March 2014	Draft scope consultation and committee recruitment
12 January 2014	First draft of the scope
23 October 2014	Recruitment for the guideline development group chair
09 October 2014	Recruit a chair for the guideline development group
04 September 2014	Update of the 2009 guideline on depression in adults

in a marginal population (drug addicts, homeless individuals and prisoners) and were based mainly on the provision of incentives. Study quality was generally low. The authors concluded that great variability exists in the studies comparing strategies for identifying interventions, objectives, and effects. The designs carried out generally have methodological deficits.

10   From Research to Practice

10.1   Primary Intention

Many NPIs are now recommended as primary intention treatments. For example, the clinical guidelines committee of the American College of Physicians recommends for acute low back pain the first line of treatment as local heat, therapeutic massage, acupuncture, or spinal manipulations, and for the second line, if there is no improvement, nonsteroidal anti-inflammatory drugs (NSAIDs) or myorelaxants (Qaseem et al. 2017). In case of chronic low back pain, the first line is exercise, multidisciplinary rehabilitation, acupuncture, meditation, tai chi, yoga, relaxation, biofeedback, or spinal manipulation; the second line, NSAIDs with or without tramadol, duloxetine, or opioids (Qaseem et al. 2017).

A mind-body program is suggested for autoimmune disorders such as chronic fatigue syndrome, rheumatoid arthritis, and inflammatory bowel disease (Rakel 2018). Oral administration of the herb feverfew, up to 125 mg/day of the dried leaf, may be beneficial for headache (Rakel 2018).

10.2   NPIs Resolutely Complement Biomedical Treatments

Creating new effective drugs is not enough. Patients need to take them. The WHO estimates that 50% of chronically ill patients with chronic disease do not take their medication as prescribed by their doctor (WHO 2003). NPIs like electronic pill boxes or disease education programs improve the proper use of drugs, and therefore their effectiveness in real life. A 1-month therapeutic education program allows for considerable savings, for example, €481 per year per patient in the case of chronic obstructive pulmonary disease (Ninot et al. 2011). Conditions most solidly backed up by evidence for the use of acupuncture methods are chemotherapy-induced nausea/vomiting, postoperative nausea/vomiting, and idiopathic headache (Ernst 2009).

10.3   After Biomedical Treatments

An adapted physical activity program helps reduce the risk of prostate cancer recurrence by 38% (Friedenreich et al. 2016) and the risk of recurrence of many other cancers as well (McTiernan et al. 2019). Many NPIs have significant benefits after a period of hospitalization.

10.4   The Danger of the Internet

It is essential not to make hasty connections between an NPI and a health problem. Just because an NPI works for a disease does not mean that it should be generalized to broader problems. For now, it is important for those wishing to use NPIs to be careful about the advice given by websites sponsored by advertisements, even if it is claimed to be offered by a doctor. It is safer not to use Wikipedia as a means of self-prescribing, particularly in light of lobbying and various other influences. Health is far too precious.

Table 4.6 Targeting health goal in NPIs

General goal	Examples used in NPIs
Prevent	Avoid a health problem Enhance an aptitude or capacity
Care	Reduce a symptom Complete biological treatment
Cure	Cure a health problem

Table 4.7 Main outcomes assessment in intervention studies

Evaluated outcomes	Examples used in NPI
Clinical symptoms	Fatigue, pain, fever, dyspnea, trembling, hypertension, agnosia, alopecia, amenorrhea, atrophy, walking abnormality, anorexia, apraxia, constipation
Behavioral	Physical inactivity, smoking, alcoholism, inappropriate diet, risky sexual behavior, violent behavior, non-compliance with prescribed treatment
Biological	Blood sugar level, hypertension
Psychological	Health-related quality of life, well-being, self-esteem, anxiety, depression, cognitive performance
Social	Social participation, work
Economic	Direct cost, hospitalization, indirect care expenses, social assistance, duration before return to work
Epidemiological	Survival, disease-free survival, lifespan without disability, lifespan without recurrence

10.5 How to Find High-Quality Information on NPIs?

This book encourages anyone contemplating the use of an NPI not to trust websites and magazines underlining the merits of a particular NPI that do not cite the exact reference source for the study and meta-analysis that assessed its effectiveness.

A bibliographic search for a clinical trial or intervention study must use the right keywords in English and clearly define the health objective targeted (Table 4.6).

Bibliographic queries must pay attention to the efficiency markers chosen; they are more and more numerous. Patients pay more attention to psychological and symptomatic markers, while doctors pay more attention to biological markers, and decision makers to socioeconomic markers (Table 4.7).

Once the pdf article is found, several points must be checked before reading it in detail and considering it as relevant:

- Prior registration of the study protocol in an official database (e.g., ClinicalTrials, Chinese Clinical Trial Register)
- Prior approbation of the protocol design by a qualitative ethics committee (e.g., National Committee and not local)

- Mention of links/conflicts of interest (e.g., sponsors, promotors)
- Trial citations in a systematic review or a meta-analysis (e.g., Cochrane)
- Contribution of an academic institution (e.g., university)

11 Conclusion

The rise of rigorous clinical trials assessing NPIs since 2010 has accelerated the determination of relevant indications and practices. Decision makers and users now have evidence-based syntheses and meta-syntheses to implement specific NPIs and compare their benefits and risks. Considered as slow-care strategies, the majority of NPIs do not cure serious illnesses and acute diseases. They alleviate symptoms and improve the quality of life. They sustainably change health behaviors. They improve healthy aging. They prevent and delay the onset of chronic diseases. Some are cost-effective. Researchers are encouraged to continue to develop new clinical trials, following the current recommendations that minimize the risk of bias, and all of these with a sample size that is adequate for its objective.

Key Points

- The multiplication of rigorous intervention studies and clinical trials on NPIs since 2010 has accelerated the determination of relevant indications and good practice dissemination.
- Systematic reviews have improved the understanding, the comparison, and the implementation of NPIs.
- Considered as slow-care treatments, NPIs do not cure serious and acute illnesses.
- NPIs improve and alleviate symptoms, and improve quality of life.
- NPIs improve health behaviors sustainably.
- NPIs participate in healthy aging strategies.
- NPIs complement authorized biomedical treatments.
- NPIs prevent and delay the onset of chronic diseases and unplanned hospitalizations.
- NPIs are cost-effective in many cases.
- An exhaustive inventory of NPIs that are efficient, as well as those that are ineffective and dangerous, is underway, but sufficient studies and a general classification are still lacking.

References

Al-Toma, A., Volta, U., Auricchio, R., et al. (2019). European Society for the Study of Coeliac Disease (ESsCD) guideline for coeliac disease and other gluten-related disorders. *United European Gastroenterology Journal, 7*(5), 583–613. https://doi.org/10.1177/2050640619844125.

Amraoui, J., Pouliquen, C., Fraisse, J., Dubourdieu, J., Rey Dit Guzer, S., Leclerc, G., de Forges, H., Jarlier, M., Gutowski, M., Bleuse, J. P., Janiszewski, C., Diaz, J., & Cuvillon, P. (2018). Effects of a hypnosis session before general anesthesia on postoperative outcomes in patients who underwent minor breast cancer surgery: The HYPNOSEIN randomized clinical trial. *JAMA Network Open, 1*(4), e181164. https://doi.org/10.1001/jamanetworkopen.2018.1164.

Anderson, J. G., Lopez, R. P., Rose, K. M., & Specht, J. K. (2017). Nonpharmacological strategies for patients with earlystage dementia or mild cognitive impairment: A 10Year update. *Research in Gerontological Nursing, 10*(1), 5–11. https://doi.org/10.3928/19404921-20161209-05.

Apaydin, E. A., Maher, A. R., Shanman, R., et al. (2016). A systematic review of St. John's wort for major depressive disorder. *Systematic Reviews, 5*(1), 148. https://doi.org/10.1186/s13643-016-0325-2.

Ballard, C., Khan, Z., Clack, H., & Corbett, A. (2011). Nonpharmacological treatment of Alzheimer disease. *Canadian Journal of Psychiatry, 56*(10), 589–595. https://doi.org/10.1177/070674371105601004.

Burckhardt, M., Herke, M., Wustmann, T., Watzke, S., Langer, G., & Fink, A. (2016). Omega-3 fatty acids for the treatment of dementia. *Cochrane Database of Systematic Reviews, 4*, CD009002. https://doi.org/10.1002/14651858.CD009002.pub3.

California Workgroup on Guidelines for Alzheimer's Disease Management. (2018). *Guideline for Alzheimer's Disease management, final report*. Los Angeles: State of California.

Cammisuli, D. M., Danti, S., Bosinelli, F., & Cipriani, G. (2016). Non-pharmacological interventions for people with Alzheimer's Disease: A critical review of the scientific literature from the last ten years. *European Geriatric Medicine, 7*, 57–64. https://doi.org/10.1016/j.eurger.2016.01.002.

Campbell, A. J., & Robertson, M. C. (2007). Rethinking individual and community fall prevention strategies: A meta-regression comparing single and multifactorial interventions. *Age and Ageing, 36*(6), 656–662. https://doi.org/10.1093/ageing/afm122.

Campbell, A. J., Robertson, M. C., Gardner, M. M., Norton, R. N., Tilyard, M. W., & Buchner, D. M. (1997). Randomised controlled trial of a general practice programme of home based exercise to prevent falls in elderly women. *British Medical Journal, 315*(7115), 1065–1069. https://doi.org/10.1136/bmj.315.7115.1065.

Carayol, M., Bernard, P., Boiche, J., Riou, F., Mercier, B., Cousson-Gelie, F., Romain, A. J., Delpierre, C., & Ninot, G. (2013). Psychological effect of exercise in women with breast cancer receiving adjuvant therapy: What is the optimal dose needed? *Annals of Oncology, 24*(2), 291–300. https://doi.org/10.1093/annonc/mds342.

Carayol, M., Ninot, G., Senesse, P., Bleuse, J. P., Gourgou, S., Sancho-Garnier, H., Sari, C., Romieu, I., Romieu, G., & Jacot, W. (2019). Short- and long-term impact of adapted physical activity and diet counseling during adjuvant breast cancer therapy: the "APAD1" randomized controlled trial. *BMC Cancer, 19*(1), 737. https://doi.org/10.1186/s12885-019-5896-6.

Chou, R., Deyo, R., Friedly, J., Skelly, A., Weimer, M., Fu, R., Dana, T., Kraegel, P., Griffin, J., & Grusing, S. (2017). Nonpharmacologic therapies for low back pain: A systematic review for an American College of Physicians clinical practice guideline. *Annals of Internal Medicine, 166*(7), 493–505. https://doi.org/10.7326/M16-2459.

Cohen-Mansfield, J. (2013). Nonpharmacologic treatment of behavioral disorders in dementia. *Current Treatment Options in Neurology, 15*, 765–785. https://doi.org/10.1007/s11940-013-0257-2.

Cuijpers, P., Berking, M., Andersson, G., Quigley, L., Kleiboer, A., & Dobson, K. S. (2013). A meta-analysis of cognitive-behavioural therapy for adult depression, alone and in comparison with other treatments. *Canadian Journal of Psychiatry, 58*(7), 376–385. https://doi.org/10.1177/070674371305800702.

Cuijpers, P., Karyotaki, E., Eckshtain, D., Ng, M. Y., Corteselli, K. A., Noma, H., Quero, S., & Weisz, J. R. (2020). Psychotherapy for depression across different age groups: A systematic review and meta-analysis. *JAMA Psychiatry, 18*, e200164. https://doi.org/10.1001/jamapsychiatry.2020.0164.

Ernst, E. (2009). Acupuncture: what does the most reliable evidence tell us? *Journal of Pain and Symptom Management, 37*(4), 709–714. https://doi.org/10.1016/j.jpainsymman.2008.04.009.

Ernst, E., Pittler, M. H., Wider, B., & Boddy, K. (2008). *Oxford handbook of complementary medicine*. Oxford: Oxford University Press.

French High Authority for Health (2011). *Développement de la prescription de thérapeutiques non médicamenteuses validées*. Paris: HAS.

French High Authority for Health (2018). *Parcours de soins des patients présentant un trouble neurocognitif associé à la maladie d'Alheimer ou à une maladie apparentée*. Paris: HAS.

Friedenreich, C. M., Wang, Q., Neilson, H. K., Kopciuk, K. A., McGregor, S. E., & Courneya, K. S. (2016). Physical activity and survival after prostate cancer. *European Urology, 70*(4), 576–585. https://doi.org/10.1016/j.eururo.2015.12.032.

Gueguen, J., Barry, C., Seegers, V., & Falisard, B. (2013). *Évaluation de l'efficacité de la pratique de l'auriculothérapie*. Paris: INSERM.

Haller, H., Winkler, M. M., Klose, P., Dobos, G., Kümmel, S., & Cramer, H. (2017). Mindfulness-based interventions for women with breast cancer: an updated systematic review and meta-analysis. *Acta Oncologia, 56*(12), 1665–1676. https://doi.org/10.1080/0284186X.2017.1342862.

Holvast, F., Massoudi, B., Oude Voshaar, R. C., & Verhaak, P. F. M. (2017). Non-pharmacological treatment for depressed older patients in primary care: A systematic review and meta-analysis. *PLoS One, 12*(9), e0184666. https://doi.org/10.1371/journal.pone.0184666.

Horneber, M., Bueschel, G., Dennert, G., Less, D., Ritter, E., & Zwahlen, M. (2012). How many cancer patients use complementary and alternative medicine: A systematic review and metaanalysis. *Integrative Cancer Therapies, 11*(3), 187–203. https://doi.org/10.1177/1534735411423920.

Karssemeijer, E. G., Aaronson, J. A., Bossers, W. J., Smits, T., Olde Rikkert, M. G. M., & Kessels, R. P. C. (2017). Positive effects of combined cognitive and physical exercise training on cognitive function in older adults with mild cognitive impairment or dementia: A metaanalysis. *Ageing Research Reviews, 40*, 75–83. https://doi.org/10.1016/j.arr.2017.09.003.

Lee, M. S., Pittler, M. H., Guo, R., & Ernst, E. (2007). Qigong for hypertension: A systematic review of randomized clinical trials. *Journal of Hypertension, 25*(8), 1525–1532. https://doi.org/10.1097/HJH.0b013e328092ee18.

Linde, K., Berner, M. M., & Kriston, L. (2008). St John's Wort for major depression. *Cochrane Database of Systematic Reviews, 4*, CD000448.

Linton, S. J., Boersma, K., Traczyk, M., Shaw, W., & Nicholas, M. (2015). Early workplace communication and problem solving to prevent back disability: Results of a randomized controlled trial among high-risk workers and their supervisors. *Journal of Occupational Rehabilitation, 26*(2), 150–159. https://doi.org/10.1007/s10926-015-9596-z.

McTiernan, A., Friedenreich, C. M., Katzmarzyk, P. T., Kenneth, E., Powell, K. E., Macko, R., Buchner, D., Pescatello, L. S., Bloodgood, B., Tennant, B., Vaux-Bjerke, A., George, S. M., Troiano, R. P., & Piercy, K. L. (2019). Physical activity in cancer prevention and survival: A systematic review. *Medicine and Science in Sports and Exercise, 51*(6), 1252–1261. https://doi.org/10.1249/MSS.0000000000001937.

Micozzi, M. S. (2019). *Fundamentals of complementary, alternative and integrative medicine* (6th ed.). St. Louis: Elsevier.

NICE. (2009). *Depression in adults: Recognition and management.* London: National Institute of Excellence for Health and Care Excellence.

Nickel, F., Barth, J., & Kolominsky-Rabas, P. L. (2018). Health economic evaluations of non-pharmacological interventions for persons with dementia and their informal caregivers: A systematic review. *BMC Geriatrics, 18*(1), 69. https://doi.org/10.1186/s12877-018-0751-1.

Ninot, G., Moullec, G., Picot, M. C., Jaussent, A., Hayot, M., Desplan, J., Brun, J. F., Mercier, J., & Prefaut, C. (2011). Cost-saving effect of supervised exercise associated to COPD self-management education program. *Respiratory Medicine, 105*, 377–385. https://doi.org/10.1016/j.rmed.2010.10.002.

Qaseem, A., Wilt, T. J., McLean, R. M., & Forciea, M. A. (2017). Clinical guidelines committee of the American College of Physicians. Noninvasive treatments for acute, subacute, and chronic low back pain: A clinical practice guideline from the American College of Physicians. *Annals of Internal Medicine, 166*(7), 514–530. https://doi.org/10.7326/M16-2367.

Rahani, V. K., Vard, A., & Najafi, M. (2018). Claustrophobia game: Design and development of a new virtual reality game for treatment of claustrophobia. *Journal of Medical Signals and Sensors, 8*(4), 231–237. https://doi.org/10.4103/jmss.JMSS_27_18.

Rakel, D. (2018). *Integrative medicine* (4th ed.). Philadelphia: Elsevier.

Riquelme-Miralles, D., Palazón-Bru, A., Sepehri, A., & Gil-Guillén, V. F. (2019). A systematic review of non-pharmacological interventions to improve therapeutic adherence in tuberculosis. *Heart & Lung, 48*, 452–461. https://doi.org/10.1016/j.hrtlng.2019.05.001.

Robertson, M. C., Devlin, N., Scuffham, P., Gardner, M. M., Buchner, D. M., & Campbell, A. J. (2001). Economic evaluation of a community based exercise programme to prevent falls. *Journal of Epidemiology and Community Health, 55*(8), 600–606. https://doi.org/10.1136/jech.55.8.600.

UsAgainstAlzheimer's. (2019). *Non-pharmacological therapies in Alzheimer's disease: A systematic review.* Washington, DC: UsAgainstAlzheimer's.

Van Der Kolk, B. A., Spinazzola, J., Blaustein, M. E., Hopper, J. W., Hopper, E. K., Korn, D. L., & Simpson, W. B. (2007). A randomized clinical trial of eye movement desensitization and reprocessing (EMDR), fluoxetine, and pill placebo in the treatment of posttraumatic stress disorder: Treatment effects and long-term maintenance. *Journal of Clinical Psychiatry, 68*(1), 37–46. https://doi.org/10.4088/jcp.v68n0105.

Van der Steen, J. T., Smaling, H. J., van der Wouden, J. C., Bruinsma, M. S., Scholten, R. J., & Vink, A. C. (2018). Music-based therapeutic interventions for people with dementia. *Cochrane Database of Systematic Reviews, 7*(7), CD003477. https://doi.org/10.1002/14651858.CD003477.pub4.

Vellas, B., Coley, N., Ousset, P. J., Berrut, G., Dartigues, J. F., Dubois, B., Grandjean, H., Pasquier, F., Piette, F., Robert, P., Touchon, J., Garnier, P., Mathiex-Fortunet, H., & Andrieu, S. (2012). Long-term use of standardized ginkgo biloba extract for the prevention of Alzheimer's disease (GuidAge): A randomized placebo-controlled trial. *Lancet Neurology, 11*, 851–859. https://doi.org/10.1016/S1474-4422(12)70206-5.

Volkow, N. D., & McLellan, A. T. (2016). Opioid abuse in chronic pain – Misconceptions and mitigation strategies. *New England Journal of Medicine, 374*(13), 1253–1263. https://doi.org/10.1056/NEJMra1507771.

Wang, C., Schmid, C. H., Rones, R., Kalish, R., Yinh, J., Goldenberg, D. L., Lee, Y., & McAlindon, T. (2010). A randomized trial of tai chi for fibromyalgia. *New England Journal of Medicine, 363*(8), 743–754. https://doi.org/10.1056/NEJMoa0912611.

WHO. (2003). *Adherence to long-term therapies: Evidence for action.* Geneva: WHO.

Wieland, L. S., Skoetz, N., Pilkington, K., Vempati, R., D'Adamo, C. R., & Berman, B. M. (2017). Yoga treatment for chronic non-specific low back pain. *Cochrane Database of Systematic Reviews, 1*, CD010671. https://doi.org/10.1002/14651858.CD010671.pub2.

Williams, N. H., Wilkinson, C., Russel, I., Edwards, R. T., Hibbs, R., Linck, P., & Muntz, R. (2003). Randomized osteopathic manipulation study (ROMANS): Pragmatic trial for spinal pain in primary care. *Family Practice, 20*(6), 662–629. https://doi.org/10.1093/fampra/cmg607.

Woods, B., Manca, A., Weatherly, H., Saramago, P., Sideris, E., Giannopoulou, C., Rice, S., Corbett, M., Vickers, A., Bowes, M., MacPherson, H., & Sculpher, M. (2017). Cost-effectiveness of adjunct non-pharmacological interventions for osteoarthritis of the knee. *PLoS One, 12*(3), e0172749. https://doi.org/10.1371/journal.pone.0172749.

World Alzheimer Report. (2018). *World Alzheimer Report 2018. The state of the art of dementia research: New frontiers.* London: Alzheimer's Disease International.

Yorozuya, K., Kubo, Y., Tomiyama, N., Yamane, S., & Hanaoka, H. (2019). A systematic review of multimodal non-pharmacological interventions for cognitive function in older people with dementia in nursing homes. *Dementia and Geriatric Cognitive Disorders, 48*(1–2), 1–16. https://doi.org/10.1159/000503445.

Chapter 5
The Dangers of Non-pharmacological Interventions

1 Introduction

Natural remedies and traditional medicines are supposed to be safe. However, every health solution presents risks, provokes side effects, and stimulates nocebo processes. Without enough research, surveillance, regulation, and ethics, the complementary medicine area looks like a jungle. The best meets the worst, the useless meets the abusive. This is all the more serious when these practices are aimed at vulnerable, fragile, docile, gullible, sick people, and/or those who are ready to make any sacrifice to achieve their goal. This chapter presents the main risks to be avoided.

2 Cognitive Biases and Logical Fallacies in a Globalized Disinformed World

Understanding and following instructions in the notice for a drug are clear and easy. Each user can identify in a formalized and a standardized document all product details: pharmacotherapeutic category, indications, contraindications, precautions for use, adverse effects, and production process. The dosage is explicitly stated. These criteria have been regulated, for example, by the U.S. Food and Drug Administration (FDA) in the United States or by the European Medicines Agency (EMA) in Europe.

In alternative medicine, there is no equivalent. No document is available. No standardized notice for the patient. No manualized specifications for the professional. No guarantee. Few practitioners are able to describe their methods precisely. These absences leave space for a lot of speculation, for preconceived ideas and beliefs, and, therefore, for all kinds of excesses and amalgams.

2.1 Marketing and Media

Advertising, disguised promotion including through social media, is everywhere, playing with words, recommending products, disseminating miraculous healings. Medical terms are replaced by common words in documents, posts, and videos. For example, the word fatigue replaces the medical word asthenia, the word stress with that of anxiety. The terms of clinical research are diverted: "efficacy" becomes "efficient," "study" becomes "survey," "trial" becomes "test," and "validated" becomes "approved by experts".

The packaging, the places of practice, and the clothes of professionals lead to a belief in what they are not. Asterisks refer to information written in such small characters that nobody can read it. The Internet is then there to over-inform, to misinform, in favor of questionable businesses and more or less malicious people.

Pseudo-scientific and pseudo-clinical arguments are easily found on the Internet. The testimony of a patient on television is often more impactful than an expert who describes the complexity of a phenomenon (e.g., the example of forest-bathing in Chap. 1). Humankind has a thirst for unlimited knowledge. The only problem is that information is so readily available today, so easily accessible, and presented on the same level, meaning that anything can be stated as well as its opposite. There will always be an expert, a great witness, or a media personality to contradict even the most absolute truths. Everyone ends up forging their opinion from data that corroborates their intuitions. Journalists like controversies to create a buzz and get an audience on TV or radio, or in print media, but this can come with the parallel increase in doubts and conspiracy-theory types of ideas. All opinions are worthwhile at the end of the day; all information is important to state, true or false.

Goop Lab on Netflix Since 2020
The lifestyle brand of Gwyneth Paltrow intends to provide wellness and health advice. The line between science and pseudoscience has become blurred with the oxymoron title of the TV series on Netflix, the *"Goop Lab."* Qualified scientists speak alongside self-declared experts. Facts are presented alongside personal theories or opinions.

2.2 In the Name of Freedom

The deregulation of the information market in the name of freedom starts from a commendable intention. Who could oppose free access to information except a few dictators? New technologies have dramatically increased the amount of information that is circulating and available. In 2 years, the information produced now is equivalent to the total of that produced since the Neolithic (12,000 BC). Humans are drowning in a colossal mass of information that prevents them from distinguishing between different items correctly. Artificial intelligence systems and data mining algorithms perform poorly and throw up a significant rate of false positives. Automating the process of sorting or distinguishing seems difficult in the face of such a mass of data. Indeed, any sorting, algorithmic or human, could be charged with limiting freedom of expression. Pressure groups take advantage of this weakness to advocate community causes and sometimes to vehemently defend minority opinions. These tyrannies of minorities can then occupy the headlines of the media scene.

In the name of freedom of expression, digital information systems such as search engines have given priority to social visibility; in other words, to popularity (e.g., number of clicks, number of views, number of likes, satisfaction rates) rather than to scientific validity. While Facebook pays attention to conspiracy theories, freedom of individual expression remains its watchword. The popularity of information prevails over its veracity. Truths are transformed, popular beliefs become realities. A testimony, a photo, or a video becomes proof.

2.3 Three Main Logical Fallacies in Favor of Alternative Medicines

Collective beliefs about alternative medicines and ancestral health practices become individual truths under the influence of erroneous reasoning. A logical fallacy stems from an error in a logical argument.

The first logical fallacy, the intuition bias, is based on the fact that humans need to believe despite the general level of study that is increasing, despite the availability of information, despite our so-called Cartesian societies. The human brain is tempted to follow intuition at all times. There are promises that the person wants to believe, and even more so in the case of individual or collective fear such as in the face of a risk of illness or real illness. A conspiracy theory goes in the direction of our intuitions; it only confirms what we imagined. When science is groping and slow to respond, when experts are opposed on TV or elsewhere, the brain takes the step of going in the direction of intuition. Uncertainty strengthens this choice. A breach of trust in science takes place and legitimizes conspiracy theories (e.g., debates around hydroxychloroquine during the pandemic of COVID-19).

Hydroxychloroquine: What a Story

Did you hear about this drug before COVID-19? An internationally renowned medical professor told a journalist that the treatment of hydroxychloroquine is working to treat patients with COVID-19. During the apex of the pandemic, knowing that this treatment does not have many side effects, that it is available, and that it is not expensive, even if there was some doubt, why was it not applied? A president of a large country recommended the drug as a miracle. But many experts refused to disseminate this drug widely without evidence from rigorous research. These refusals opened the door to a conspiracy theory. Malicious and highly placed people were deemed to have prevented this professor from speaking. They hide things. It is the province against the capital. It is the people against the elite. Then, as if to make matters worse, the "LancetGate" case arises with the approximative big data study published online ahead of print on 22 May 2020 and retracted on 5 June 2020 (Mehra et al. 2020). What if David was right against Goliath? The total absurdity was reached when a survey was commissioned by a French media outlet to settle the scientific debate on the efficacy of hydroxychloroquine. You can imagine what can happen with an alternative medicine.

The second logical fallacy is the causality bias. A correlation becomes a causality. A co-occurrence of facts must have a causal link at all costs (e.g., 5G and COVID-19). The ancient post hoc, *ergo propter hoc* fallacy ("it worked for me") is engraved into the human mind. *"If a patient receives a treatment and then gets better, what could be more logical than to assume that the treatment was the cause of the improvement? This conclusion seems as obvious to patients – and many clinicians – as it is fallacious. Proponents of alternative medicine employ this fallacy incessantly to convince us that ineffective treatments are, in fact, effective"* (Ernst 2013, p. 1025). Subsequently is not the same as consequently. Causal inferences based on anecdotes are, therefore, highly problematic and certainly not a sound basis for robust conclusions about the efficacy of treatments.

5G and COVID-19

A conspiracy theory claimed that 5G technology can spread the coronavirus, COVID-19. The myth gained in popularity when a Belgian doctor linked the "dangers of 5G" to the virus during an interview in January 2020. A video of an American doctor published on March 18, 2020 and viewed more than 668,000 times claimed that Africa is less affected by COVID-19 because it is not a 5G region. Stop5G Australia members have claimed the Ruby Princess cruise liner's link to 600 reported infections and 11 deaths is because cruises are "radiation saturated." But, the cruise passengers did not have 5G services on board. The technology supposedly negatively affects the immune system. Conspiracy theories have motivated arson attacks on 5G towers in Belfast, Liverpool, and Birmingham [note: as well as in a number of countries other than the United Kingdom]. To many of us, it is obvious a human virus cannot spread via radio signals, and such a conspiracy may be linked to a wider distrust of the government in general, but, at a time when millions are relying on fast Internet, myriad groups and public figures continue to perpetuate it.

The third logical fallacy is about the claim of lack or absence of evidence. The absence of evidence is not evidence of the absence of effect. No or no good evidence for the effectiveness of a treatment cannot assume it is non-effective. However, to argue that it is relevant to use this treatment until evidence emerges that proves it to be ineffective is a real danger. In health care, it is unwise and arguably unethical to give the benefit of the doubt to under-researched therapies. *"Science certainly has its limits. Yet, when it comes to testing therapeutic claims, it provides us with fairly adequate tools to assess them. Even if the claim is that a particular holistic, individualized and complex form of energy healing makes you feel better, live longer or experience life more wholesomely, the hypothesis is scientifically testable. Even if no validated outcome measure exists for a particular claim, scientists should be able to develop one. The notion that "a therapy defies scientific testing" merely discloses a lack of understanding of what science can achieve."* (Ernst 2013, p. 1025).

CAM vs. NPI

Complementary and alternative medicine (CAM) suffers from the influences of postmodernism, anti-science, and regressive thinking, as Edzard Ernst (2009) likes to point out. You might as well proscribe the term of alternative medicine and use the term NPI, which relates to evidence-based methods of prevention or care.

2.4 Two Main Cognitive Biases in Favor of Alternative Medicines

A cognitive bias is a systematic error in thinking that occurs when people are processing and interpreting information in the world around them, affecting their decisions and judgments. A cognitive bias is rooted in thought processing errors often arising from problems with memory, attention, attribution, and other mental mistakes.

The first cognitive bias is the negativity bias. Our brains have been wired and selected for generations to pay attention to threat and risk. The brain is more sensitive to negative information. Negative information is received and stored at three times the rate of positive information. A train that arrives late makes the headlines, never one that arrives on time. Fears and threats are increasing. Fear causes people to become hypochondriacal. We no longer rationalize fears. They are contagious; they become norms. Other negative emotions are shared as well as fear, anger, and indignation. Alternative medicine actors play on fear, anger, and indignation.

The second cognitive bias relates to saving time and effort and is called the Dunning-Kruger effect (Kruger and Dunning 1999). Incompetent people tend to overestimate their own skill levels, fail to recognize the genuine skill and expertise of other people, and fail to recognize their own mistakes and lack of skill (Fig. 5.1). It relates to an effect of overconfidence according to which the least qualified in an area overestimates his or her competence. Reading one or two articles about medicine is enough to give the impression of knowing. That deficit in skill and knowledge creates two consequences. The individual performs poorly in the domain in which he or she is incompetent. He or she is unable to recognize his or her own mistakes. To err is human. But, to confidently persist in erring is just stupid.

2.5 Scientific Truth Is Counterintuitive

FakeMed conspiracy theories and ideological opportunism make social networks happy. True or false, it does not matter; they fuel exchanges by taking advantage of the opportunity to slip in some advertising messages. News and fake news circulate

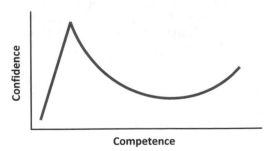

Fig. 5.1 The Dunning-Kruger effect relates to overconfidence according to which the least qualified in an area overestimates his or her competence (Kruger and Dunning 1999)

very quickly, without borders, 24 h a day, 7 days a week. FakeMed sellers take advantage of it.

FakeMed Movement
The petition of 174 French health professionals against alternative medicines published in the French newspaper *FigaroVox* on 18 March 2018 was excessive because it mixed alternative medicines and evidence-based NPIs but has had the merit of alerting opinion on the complex topic. In 2020, the website of the structured FakeMed movement mentioned that alternative medicines *"are ineffective beyond any placebo effect and can even prove to be dangerous. They can be dangerous because they treat irrelevant symptoms and over-medicalize populations, giving the illusion that any situation can be solved with a "treatment". They can be dangerous because they fuel and rely on a fundamental distrust of conventional medicine as shown by the unjustified polemics surrounding vaccines. Finally, they can be dangerous because their use delays diagnoses and necessary treatments, sometimes leading to dramatic consequences, especially in the treatment of serious diseases such as cancers"* (French FakeMed website 2020).

In the vast competition that is organized on the information market, in the face of this period of hysteria on social networks and the radicalization of opinions, science will always be lagging behind. Rationality has a competitive disadvantage compared to gullibility. Rationality requires time that the instantaneity of our societies no longer allows. Rationality requires a development format that cannot be reduced to 144 characters. Scientists are on the alert; they keep coming together to fight against *Infox*. However, it takes a lot of effort to break the logic of the gullibility epidemic. Certain statistical realities are formidable to deal with and the possible errors are numerous. Our critical spirit is damaged. Science does not tell the absolute truth. It prioritizes information. There may still be people who believe that the earth is flat, but science has demonstrated the opposite and made it known to the majority of people on the planet. The problem is that scientists and teachers do not continue to teach it to future generations.

3   The Dangers of Natural Products and Traditional Practices

Ayurvedic medicine is 3,000 years old. The medicine of ancient Egypt is 2,600 years old. Chinese medicine is 2,500 years old. The medicine of ancient Greece is 2,400 years old. A medicine known since the dawn of time, which has become traditional (trivially as a home remedy, for example), has no deleterious consequences

for health. A treatment, because it is natural, because it uses natural components,
would be safe (Fig. 5.2). Its harmlessness would go without saying.

> **Reiki**
> Reiki is a manual therapy of Japanese origin founded by Mikao Usui follow-
> ing a mystical revelation that led him to receive the "keys to healing" at the
> end of the nineteenth century. The method transfers life energy into the
> patient's body that is supposed to stimulate his or her self-healing resource.
> *"The therapy lacks plausibility and its effectiveness is not supported by good
> evidence. There are no major risks. The risk/benefit balance fails to be posi-
> tive"* (Ernst and Smith 2018, p. 218).

Traditional or natural, in both cases, advocates argue that there is no reason to
want to check their health risks since these medicines are practiced by "masters"
above all suspicion, often disinterested in material things. Interrogating followers
convinced by personal experience would do nothing if not support the words of the
master. These testimonies of users, famous or not, and these singular illustrations of
professionals abound on social networks and Internet pages.

Popular Incentive

Showbiz stars like Cameron Diaz and intellectuals like Pierre Rabhi have been promoting natural methods for health. Natural health and traditional medicine events (e.g., international conferences, congresses, seminars, meetings, exhibitions) are organized every month around the planet. Chinese medicine advocates push WHO and states to integrate their own traditional practices. But health is too precious a good to recommend natural and traditional medicines without having scientific and clinical evidence of their effectiveness and above all of their harmlessness (or that they present no danger to health).

A synthesis of three systematic reviews shows that acupuncture, chelation therapy, and chiropractic have caused 100 deaths (Ernst 2011). Fasting can interfere with chemotherapy. Indeed, while the results obtained by fasting in cellular and animal models might theoretically be transferred to patients with cancer in terms of treatment response, *"toxicity and survival remain to be ascertained"* (Caccialanza et al. 2019, p. 177). A solution made up of natural elements or a practice coming from a previous time does not mean that it is necessarily safe. A natural component paradoxically can be toxic. For example, honey is produced by bees from the genus *Apis* that collect the nectar from plants or from secretions of aphids. It is a sweet and flavorful natural product, supersaturated in sugars, with high nutritive value. Besides the sugars, other minor components are present in honey, such as minerals, polyphenols, vitamins, carotenoids, amino acids, proteins, enzymes, organic acids, and volatile compounds (Miguel et al. 2017). In addition to the described probiotic effects associated with oligosaccharides, most of the reported biological properties of honey (e.g., antioxidant, antibacterial, antiviral, anti-inflammatory, anti-ulcerous, immunomodulating, vasodilative, hypotensive, anti-hypercholesterolemic, anti-browning, disinfectant, and anti-tumor), and many of its applications, may be attributed to those minor components. Honey may be contaminated by pesticides, antibiotics, heavy metals, and other toxic compounds. The presence of such compounds may be attributed not only to accidental exposure and environmental hazards but also to compounds added by beekeepers to control honeybee diseases. In addition, honey may also be contaminated with pathogens, particularly *Clostridium botulinum* and its spores (Miguel et al. 2017).

The Case of Phytotherapy
The same plant can have a variable effect due to many parameters such as the harvest season, the degree of ripening, the extraction process, the manufacturing process, the climate, the soil, and the interactions between the various products taken by the patient (Efferth and Kaina 2011). The molecules contained can differ in quality and in quantity. Plants can be poisonous, sometimes because their names have been confused. In the 1990s, Guang Fang Ji (*Aristolochia fangchi*) was confused with Han Fang Ji (*Stephania tetrandra*) due to its proximity to a name in Chinese usage. It has caused kidney failure in more than 100 patients who have had to urgently receive a transplanted kidney and have treatment for renal cancer in the medium term. Some plants have been shown to be toxic because their toxicity was not fully understood. Cases of kidney disease were discovered late in patients who had followed a treatment, a diet for weight loss, with capsules based on *Magnolia officinalis*. Some plants have been shown to be toxic because they have been contaminated with pesticides, heavy metals, microbes, radioactivity, organic solvents, chemicals, and hormones. In the United States, a study on dietetic products containing ginseng showed that 8 products out of 22 contained pesticides prohibited for sale (quintozene, hexachlorobenzene) because they are considered carcinogenic. Another study found that 10–15% of the herbal plants produced in California contained heavy metals such as mercury or arsenic. Toxicities are also observed due to interaction with drugs. This is the case, for example, between garlic and warfarin, which, when taken together, increase the antiplatelet aggregating power of warfarin to the point of causing more frequent bleeding. Plants interact with the metabolism of many drugs, such as *Piper methysticum* with cytochrome P450.

A systematic review (Izzo and Ernst 2009) points to the most dangerous interactions between St. John's Wort (*Hypericum perforatum*), which reduces plasma concentrations and/or which increases clearance, and several drugs (alprazolam, amitriptyline, atorvastatin, chlorzoxazone, ciclosporin, deisoquine, digoxin, erythromycin, fexofenadine, gliclazide, imatinib, indinavir, irinotecan, ivabradine, mephenytoin, methadone, midazolam, nifedipine, omeprazole, oral contraceptives, quazepam, simvastatin, tacrolimus, and talinololol).

Hippocrates' medical motto, *Primum non nocere* (first, do no harm) is a focal point of NPIs. Unfortunately, a lot of progress remains to be made in the sector, since the side effects and the risks of interactions are more rarely noted in non-pharmacological than pharmacological clinical trials, and even less well monitored after their market access.

4   Ecological Disaster

Traditional medicines use "raw materials" that are far from harmless, such as certain protected animal and plant species. This cannot be ignored. These practices, rooted in ancient traditions on all continents, seem to justify the intensive exploitation of natural resources.

As demand becomes stronger, the "production system" intensifies and is organized through more or less legal channels. No matter what the consequences are on the environment; the damage to flora and fauna is considerable. Plants and fungi are cultivated intensively, using products that are harmful to the soil and future consumers.

> **Massacre of Plants and Animal Species**
> Ginseng smuggling is raging in the American Appalachian mountain range. Protected specimens like ferns and orchids are disappearing from the planet. Animals are slaughtered for a piece of their anatomy, like donkey for its skin intended for Chinese medicine. The jelly from this skin, sold in a € 700 jar, is claimed to promote healthier aging. The demand is such that Kenya has become an intensive producer of donkeys. More than 400 animals are slaughtered every day in this country. The rhinoceros' horn, on the other hand, is said to have such aphrodisiac properties that Zimbabwe and Namibia trade it intensively. And that is not all; the following list is far from exhaustive: tiger bones; scorpion extracts in Africa; turtle shell extracts from South America; jaguar bones in Bolivia (Fraser 2018); seahorse and snake extracts from Asia; elephant skin in Burma; bear bile in China; pangolin scales in Indonesia; snake extract in Australia for bush medicine. And now comes the flight of animals into European and North American zoos. That is flight away from the medical trade. How far will we go for a hypothetical better health? Nobody knows. The business, on the other hand, is flourishing.

Is not it time to conduct serious clinical trials on these practices to stop this destruction and these large-scale massacres on our fragile planet, where products are moving faster to the consumer?

5   Fashion Effects

You may have had your broccoli and red fruit phase, then the lemon, ginger and turmeric one, then the spirulina one. You may have tried one of these diets that are supposed to be "revolutionary" (Table 5.1). You may have tested anti-stress nutrient capsules. You may have tried probiotics. You may have downloaded the latest health

Table 5.1 Non-exhaustive inventory of diets that have not necessarily been assessed by a clinical trial

3 h	Dash	Kg	Potato
5/2	Detox	Kosher	Pritikin
Acid base	Dukan	Lactose free	Pure vital
Anti-cholesterol	Entry	Lemon	Red fruits
Anti-diabetes	Eone	Mac Keith	Salt-free
Antoine	Exclusion	Macrobiotic	Scarsdale
Atkins	Fiber	Mayo	Seignalet
Ayurvedic	Flexitarian	Meal replacement	Separate
Belle-Plante	Fricker	Mediterranean	Shapiro
Best Life	Gluten-free	Medium Fat	Slim data
Blood groups	Glycemic index	Miami	Small jars
Bootcamp	Grapefruit	Montignac	Soup
Cabbage soup	Greenprint	Natman	Tahiti
Carbohydrate loading	High protein	Okinawa	Thonon
Chromatic	Hollywood	Online coaching	Vegan
Chrononutrition	Hypnosis	Ornish	Vegetarian
Cohen	Jenny Craig	Orsoni	Walter Longo
Cosmonaut	Karl Lagerfeld	Paleolithic	Weight watchers
Cretan	Ketogenic	Portfolio	

application offering an original stress reliever recipe. You may have bought the latest essential oil diffuser promised to be much better than its predecessors.

Marketing is constantly accelerating sales cycles. Traditional, natural and wellness methods enjoy rare impunity. They can play all the tricks of the business, ask a glamorous celebrity to extol the merits, or appear in movies or a TV series. Psychologists in the United States found that wearing specific articles of clothing had an effect on the wearer's psychology and performance (Adam and Galinsky 2012). They concluded that clothes have a symbolic meaning. Wearing an article of clothing with a specific meaning can influence one's psychological state; it is called "enclothed cognition."

Diane de Poitiers (1500–1566), a Fashion Victim Intoxicated by a Gold Drink

Gold's supposed powers of regeneration go back to antiquity. Pliny the Elder (AD 23–79) describes the preparation of two remedies using gold and their therapeutic properties (Charlier et al. 2009). In the thirteenth century, alchemists like Michael Scot, Roger Bacon, and Arnaud de Villeneuve wrote about *aurum potabile* (drinkable gold) and how to obtain it. Drinkable gold was well known in the sixteenth century French Court, and Alexandre de la Tourette dedicated his book on the subject to King Henri III. In the seventeenth century, many doctors and chemists like Jean Beguin and Christophe Glaser published gold recipes, including drinkable gold, in their chemistry manuals. Diane de Poitiers 1500–1566 gradually got intoxicated with a gold drink advised by her doctor in order to gain better health. A toxicological study of her hair revealed a gold concentration 500 times higher than the average. The gold caused thinning of her hair, anemia, brittle bones, and digestive disorders such as anorexia, nausea, vomiting, and diarrhea.

Does it seem that the time required for marketing is compatible with that of science? Unfortunately with NPIs, science will always lag behind, but it will eventually win as it did for drugs. In a few years, you will only use NPIs that have a real benefit for your health, and will abandon the dangerous, the ineffective, or those that force you into unnecessary efforts. For NPIs, science will always lag behind, but it will eventually win as it did for drugs [note this repeats the exact wording three lines above?!].

6   Risks of Secondary Effects

Any therapy can cause undesirable effects and are generally noted either in studies or in routine practice. However, given their relative novelty and the lack of standardized validation and surveillance protocols (see Chap. 8), many intervention studies for NPIs do not identify incidents (e.g., pain, nausea) or accidents (e.g., drug interference, disease) that have occurred during the administration of an NPI and/or after.

Patients who use these solutions through their own initiative are less likely to report side effects to their doctors and health authorities in the event of a failure or incident. And yet this knowledge could benefit other patients and improve professional practices if they were known.

7   Inappropriate Doses

An NPI as a dietary supplement or a plant can be harmless at one dose, beneficial at another, and fatal at another.

An exercise program with only one session every month will have no chance of improving health. The anarchic use of mobile health tools means they function more as a gadget than as a health solution. Specifications for the use of NPIs are missing most of the time. Formal manualized notices or labels are not available to clarify dose effects and strategies of personalization. Everyone, professionals and users alike, "tinker" due to complaints of undesirable or absent effects.

8   Nocebo Effect

NPIs do not only generate a placebo effect. The opposite effect, harmful to health, is called the nocebo effect. The nocebo effect can also be reproduced when the undesirable effects of a treatment occur through mimicry. The patient knows that he or she is taking a medication and subconsciously recreates the side effects suggested by a healthcare professional, a close person, or information obtained through the media or other reading material. Between 20% and 30% of healthy subjects participating in various studies experience headache, drowsiness, or nausea, theoretically without direct relationship to the intervention assessed. Anxiogenic verbal suggestions were capable of turning tactile stimuli into pain as well as low-intensity painful stimuli into high-intensity pain. This "hyperalgia" is an increase in the perception of pain that is out of proportion to the stimulation (Benedetti et al. 2011). Nocebo suggestions of a negative outcome can produce both hyperalgesic and allodynic effects.

> **Defining Nocebo Effect**
> The nocebo effect, opposite to the placebo effect, is harmful to health. A verbal suggestion is enough to produce a nocebo effect following the administration of an inert product. A study shows this from experience with different medical suggestions based on an electrical signal or a tactile contact that is supposed to be more or less painful. The healthy subjects tested transformed this instruction into pain, and the pain of low intensity became high intensity (Colloca et al. 2008).

One study showed the nocebo effect of leg pain in healthy people who were injected with remifentanil (Bingel et al. 2011). The simple announcement of a class of drug induced a nocebo effect.

Another study compared the nocebo effect of two drugs that were supposed to be an antidepressant treatment: a tricyclic placebo and a selective serotonin reuptake inhibitor (SSRI) placebo (Rief et al. 2009). Patients receiving tricyclic placebo had more symptoms than those taking SSRI placebo, including dry mouth (19.2% vs. 1.2%), vision problems (6.9% vs. 1.2%), fatigue (17.3% vs. 5.5%), and constipation (10.7% vs. 4.2%).

Finally, stress potentiates the nocebo effect. Studies show the involvement of the dopaminergic system as in the placebo effect (Benedetti et al. 2011).

Nocebo Effects with NPIs
Nocebo effects may occur following the prescription of NPIs. Unfortunately, few studies are available on this subject.

9   Sectarian Aberrations

One of the concerns in NPI is the psychological dependence of a patient on a professional. Some of these practices meet the criteria of sectarian abuses because they are the work of "therapeutic gurus" who exert a real mental hold over the sick, stripping them of their resources.

Practitioners, including doctors, offer to help someone who is suffering. Why would one resist? They seem kind, attentive, disinterested, available, compassionate, and empathic. They multiply anecdotes relating their last successes. They deliver parts of their personal history. They speak with calm, confidence, and serenity. They have a strong aura. They are totally convinced of their purpose and their theory. They have everything. At first, they offer just a few personal tips. Then, gradually, some products, like supplements, materials, or books, must be bought. Then the so-called "professional" offers his or her services by coming to your home or by organizing internships reserved for a few carefully selected "faithful." The attractiveness of these gurus is such that people have withdrawn from conventional medicine and even died prematurely. Others have been ruined, while others sexually abused. Their families have been dislocated. The phenomenon of disease has become a dream gateway for sectarian movements that take advantage of the suffering or worry of patients and their families to exercise control over them.

The practice of NPIs is not without risk, but far from it. Caution should be exercised if they are not seriously studied and recommended by "pseudo-professionals." It is important not to hesitate to report any incident or accident as soon as possible to a doctor and a competent authority. Official webpages are now created to help patients to alert authorities about FakeMed, manipulations, and adverse effects, in the interest of future patients.

Methods Above All Suspicion

Instinctotherapy (consumption only of raw foods selected on their odor) is practiced in small scattered groups. This was strongly influenced by its French guru, Guy Claude Burger, who was found guilty of rape of minors and sentenced to 15 years' imprisonment.

An Italian doctor and naturopath, Tullio Simoncini, who was struck off by Italian law, teaches a method claiming the fungal nature of cancer and advocating its treatment with baking soda. He justified the method in parallel with the inability of conventional medicine to cure cancer. A fungus that can be treated by administering baking soda by local or parenteral injection would eliminate the cancer in 3 or 4 days.

Johanna Budwig claims to attack cancer or other diseases by having the patient ingest unheated, untreated linseed essential oil and curdled milk, called Budwig cream. The followers consider the method is proven but that it would be stifled by the cancer industry. She has reportedly been nominated seven times for the Nobel Prize in medicine, but the pharmaceutical industry has opposed it. Conspiracy theory would explain everything.

The method of the German doctor, Ryke Geerd Hamer, excludes the use of conventional means to treat an illness. The Hamer method postulates that all illness is the result of intense psychological shock and unresolved psychological conflict. Hamer was sentenced in 2004 in France to 3 years in prison for fraud and complicity in the illegal practice of medicine following a complaint filed by a man whose wife with breast cancer had died as a result of the refusal of proven treatments.

10   The Constraints of Use

Doing three hatha yoga sessions per week, going to the physiotherapist twice a week or to the psychotherapist once a week, meditating for 15–20 min every day, adjusting your diet to your blood sugar with a connected health device, working on memory every day with a specialized video game, maintaining use of an electronic cigarette in order to stop smoking, following a spa treatment for 3 weeks – this is hard work, and each NPI has its constraints. They far exceed the effort needed to absorb a drug in the morning or in the evening.

NPIs are not easy. The majority of NPIs require active participation from the user. They need sustainable effort, sometimes substantial, sometimes for the medium term. Some force you to get out of your comfort zone. Some require high levels of attendance. These constraints are summarized by the term "burden." It should not be overlooked, especially if several NPIs must be combined in the prevention or care strategy.

NPI Consumerism
Sustainable commitment is one of the keys to NPIs, so it is necessary to choose them in consideration of the risk/benefit balance, preference, availability, and personal meaning. Falling into consumerism will not make sense.

Beyond the effort and the burden generated, brakes on these activities can be added such as costs, travel times, or the stigmatization of certain practices (e.g., unknown oriental practices, exotic plants, regimes decried in the media).

11 Be Careful

11.1 Quality and Origins of an NPI Component

A food supplement does not meet the same manufacturing requirements as a drug. An essential oil does not undergo as much control over its composition as a drug. The origin of the ingredients, their quality, their storage, and their packaging are all important aspects that need to be considered without mentioning counterfeiting.

The assembly of technologies in health-connected objects is neither verified nor monitored to the same standard as that of implantable medical devices. Likewise, the reliability of an ergonomic health object is not as high as that of medical devices.

Advice
Systematically check product ingredients and official labels of manufacturing for a non-pharmacological remedy. In case of doubt, do not settle for the Internet; ask an expert questions before using it.

11.2 Self-Prescription

In medicine, diagnosis is essential. The American TV series *House* illustrated this perfectly. Finding and understanding the origin of a patient's illness is crucial. Avoiding or missing an organic, structural, or functional problem is crucial. NPIs are only health solutions.

Delaying a diagnosis by taking an alternative medicine or a natural remedy in the hope that the disease will go away over time is dangerous. Only a doctor is authorized to make this diagnosis. Healers or energeticians have abused the weakness of people by diverting them from their treating physicians. They have caused dramatic consequences for some patients when they were suffering from mild problems, which can easily be treated with surgery or conventional biological treatment (Ernst and Smith 2018).

Advice

Avoiding a consultation with a doctor and instead using self-prescribed remedies can have dramatic consequences for health. The intensity of pain or other symptoms tells nothing about the severity of a possible underlying disease. A conventional medical doctor will prescribe relevant and effective NPIs according to the diagnosis of your condition. Do not hide the use of NPIs or alternative medicine from your healthcare professional in any event.

11.3   Refusal of Conventional Biomedical Therapy

A study by Johnson et al. published in the *Journal of the American Medical Association* hit the headlines in 2018. It evaluated 190,815 Americans who were diagnosed with one of the four most common cancers (breast, prostate, lung, colorectal) between January 1, 2004 and December 31, 2013. Patients with metastatic cancer at the time of diagnosis, stage IV, who were in palliative care or under unknown treatment, were not included. All patients had to have undergone at least one conventional treatment (chemotherapy, radiotherapy, surgery, and/or hormone treatment). Patients who reported receiving *"other-unproven: cancer treatment administered by non-medical personnel"* (CAM group) were distinguished from those who did not. The analysis focused on the impact of CAM on treatment adherence and survival. The results indicate that the use of CAM was associated with the refusal of conventional anti-cancer treatments and a risk of death two times higher than that of patients not using CAM. The authors concluded that patients who received additional treatment were more likely to refuse other conventional cancer treatment and had a higher risk of death than without CAM. The difference in survival could also be explained by adherence or otherwise by all of the recommended conventional cancer treatments.

Advice

The use of an NPI must not distance the patient from a conventional medical doctor. If so, it is no longer an NPI but an alternative medicine. The most prudent thing to do is to call a qualified health professional, for example, in a university hospital center.

11.4   Professionals Left to Their Own Intuition

No consensus is currently available for the integration of NPIs into individual health pathways. Health professionals are left to fend for themselves and their own training initiative. This leaves room for countless speculations and misconceptions.

Decision makers thus think that an NPI is obvious (e.g., "it does not require prior training"), non-technical (e.g., "it cannot be codified"), modular (e.g., "they are interchangeable"), not sustainable (e.g., "it gives rise to numerous drop-outs and anarchic uses"), and adjustable to local resources (e.g., "according to the life place of the patient"). According to them, they would ultimately be only dietary recommendations mixed with common sense, protection from polluted environments, and precaution against toxic substances.

> **Advice**
> Keep in mind that an NPI requires the learning of new skills and knowledge in order to be proactive and to follow scrupulously the evidence-based protocols. Within this framework established by science, everyone can achieve this at their own pace. These treatments are slow and humane.

11.5 Gaps in Monitoring Usage and Protecting Patient Data

In the drug system, authorities have imposed a regulatory pharmacovigilance process. The EMA in Europe and the FDA in the United States have their own systems. Prevention and care professionals, and in particular doctors and pharmacists, must report precisely any minor or serious problem with a drug. Unfortunately, no equivalent system exists for NPIs.

Health authorities have recently created feedback procedures online for users that encourage patients to evaluate drugs and medical devices (e.g., HAS in France since 2017). Unfortunately, these initiatives have not been extended to NPIs.

Decision makers and insurers recommend the use of digital NPIs. These mHealth (mobile health) tools have reliable behavioral and health data. The whole question is to know what happens to the data collected. There is a risk regarding the non-respect of privacy and the commercial exploitation of this personal data. No regulation clearly exists for NPIs. In some cases, consumers are invited to order products online and enter a code allowing them to benefit from discounts but also to identify the professional "business introducer." Part of the turnover achieved can be transferred to these introducers in cash or in the form of gift vouchers. Certain professionals collect and store sensitive personal data of their customers – contact details, profession, health check-ups, and disorders – in dedicated software. None of them has yet taken measures to protect this data, with some even transmitting it directly to interested companies via the online platforms they make available to them.

> **Advice**
> New technologies present a chance to assess the relevance, acceptability, benefits, risks, and cost-effectiveness of NPIs. Contributing to a collective assessment of these interventions will help innovators, if, of course, it is done within a framework of respect for the security, confidentiality, and dignity of the user.

11.6   A Regulatory Vagueness

NPIs are situated between consumer products and biomedical treatments (Chap. 1).

On the one hand, consumer products are open to the market. They must only comply with manufacturing standards (e.g., CE standards in Europe). In France, the supervisory institution is the Directorate General for Competition, Consumption and Fraud Prevention (DGCCRF). They noted in 2010 that the prices of services vary a lot between professionals. In certain cases, no invoice is given to the user before the payment of the service, whereas it is an obligation as soon as the amount exceeds 25 € in France. In other cases, where an invoice is provided, it does not include all of the mandatory information. The labeling of the products sold, particularly by aromatherapists, is not complete, in particular as regards information on prices and composition of the products. When the professional travels to the consumer's home, the rules applying to distance and off-premises contracts are not always respected; in particular, the collection of the consumer's consent for the immediate execution of the service.

On the other hand, the rule for drugs and medical devices is that they must follow extremely strict manufacturing procedures, medical and scientific validation, and marketing and surveillance regulations. The manufacturer or professional must provide solid scientific and clinical evidence of efficacy and safety. He or she must obtain national (for example, the Food and Drug Administration in the United States) and supranational (for example, the European Medicines Agency) approvals. Any new therapy, be it pharmacological, surgical, radiological, genetic, or biotechnological, must demonstrate irrefutable evidence that the medical service rendered is superior to current treatments. The independent French organization, which is in charge of this validation and monitoring mission, is the National Agency for the Safety of Medicines and Health Products (ANSM).

Between these two fields of consumer products and biomedical treatments, more and more NPIs are used without any validation, production, marketing, advertising, and surveillance rules related to health. "The boundaries between conventional and unconventional care are blurred."

The same goes for information intended for users and decision makers. Professionals in the private sector are not accountable to any form of guardianship or conventional health professionals.

The way is clear to extol the merits of a fabulous connected object, a revolutionary diet, a nutrient recommended by Dr. Lamda, a herbal tea with miraculous virtues, an extraordinary massage, an incredible mushroom, etc.

Activity separations between regional organizations (for example, Regional Health Agency and Regional Agency for Youth, Sports, Education and Social Cohesion), between national agencies (for example, National Agency for the Safety of Medicines and National Agency for food, environmental and occupational health security), between ministries (for example, Sport and Health), and between high authorities (for example, High Council of Health and High Council of Public

Health) create gaps and legal vagueness in which some opportunists rush to propose ambiguous solutions with health risks that are neither measured nor monitored.

> **Advice**
> The use of "alternative medicine" can expose the patient to risks of which he or she is not aware and regulation does not yet exist. The minimum required of a professional is a fair commercial offer, an absence of illegal practice of medicine, and a guarantee of sufficient information for the consumer.

11.7 Heterogeneous Training

Practitioners are not always well trained, if at all. They rarely follow an initial academic university or health institute's course. A majority practice several disciplines. Self-employed and mostly after undertaking professional retraining, they consult individually or in a multidisciplinary office in order to pool their resources. Almost all of these professionals have followed training, but it is very variable in nature, ranging from a simple weekend (face-to-face or remote) to several years of training, before practicing.

They have no obligation of continuing education or research. They may belong to schools of thought and practice that do not argue among themselves. This is the case with reflexology, whereby private training hopes to preserve their "secrets" and to transmit their "skills" verbally and practically.

New professions are created, recycling traditional practices, combining them, in short applying them wrongly and across the board. Acting on health is not without risk. It is time to harmonize initial and continuous training in the sector, at the national level and, if possible, at the European and international levels.

> **Advice**
> Are we going to wait for a health scandal with serious cases revealed by the press to get out of the regulatory vagueness on NPIs? Why do we have to wait? Who really benefits from this lack and weakness? Certainly not the patient.

11.8 Methodological Errors and Frauds

In clinical trials, there may be benefits that are incorrectly attributed to the experimental treatment. It could come from non-intentional methodological errors such as:

- A parallel intervention identified or passed over in silence by the study participant
- A voluntary or involuntary change in diet
- An important proportion of false positives (e.g., patients diagnosed incorrectly and included in the study)
- A Hawthorne effect (where the participants feel that they have been chosen for the effect to succeed, and work hard so that the researcher's expected result is obtained)
- A regression toward the mean effect (a statistical phenomenon that is the consequence of a biased selection of patients)
- Other methodological biases (e.g., absence of analysis of subjects lost to follow-up or who dropped out, learning effect on measurement tools, use of non-validated instruments, measurement under non-standardized conditions)
- A conflict of interest (e.g., the researcher is at the same time the practitioner of the experimental NPI).

Over the last few years, the increase in errata, retracted articles, and open access science (e.g., data, supplements) is helping to guarantee more transparent studies.

Unfortunately, errors could be intentionally caused by fraud (McIntyre 2019), such as:

- Lack of ethics (e.g., gratification of positive subjects, exclusion of deviant participants)
- Use of wrong statistical analyses
- Data modification, fabrication, falsification
- Publication in a corrupt scientific journal

Advice
If you have any doubts about a method, (1) seek a second medical opinion and speak to a health professional outside the circle of influence, (2) consult a regional or national agent of the Health Ministry, and (3) check whether the proposed practice or product has seen the publication of clinical trials in scientific peer-reviewed medical journals and received positive citations from official reports (and, if possible, access to additional elements).

12   Conclusion

NPIs rarely have serious health consequences in the very short term. However, they should not delay a medical diagnosis or lead to the loss of a chance of recovery by recognized biomedical treatments. "Natural" does not mean safe for health, without side effects or without risk of interference with a biomedical treatment. Illness or pain is an ideal mode of recruitment for a sect or a guru. A holistic approach cannot pretend to solve all health problems. An NPI must have a clear health objective, precise content, and a coherent explanatory theory. The real value of a treatment is not determined by its absolute risk but by the balance between risk and benefit. If a treatment is potentially lifesaving, then substantial risks can be tolerated. If a therapy has no benefit, then even a small risk would weigh heavily and the risk-benefit balance would not be positive (Ernst 2013). Biomedical treatments and NPIs are in the same boat for better human health.

Key Points

- NPIs rarely have serious health consequences in the very short term. However, they should not delay a medical diagnosis or lead to the loss of a chance of recovery by recognized biomedical treatments.
- An NPI must have a clear health objective, precise content, and a coherent explanatory theory.
- Natural does not mean safe for health, without side effects or without risk of interference with a biomedical treatment.
- Illness or pain is an ideal mode of recruitment for a sect or guru.
- Ensure the provenance of the product and the training of the professional.
- Discuss NPIs with your doctor, especially if you are already taking a medication. Report any problems or side effects to a local or national health agency.
- If a therapy has no benefit, then even a small risk would weigh heavily, and the risk-benefit balance would not be positive.

References

Adam, H., & Galinsky, A. D. (2012). Enclothed cognition. *Journal of Experimental Social Psychology, 48*(4), 918–925. https://doi.org/10.1016/j.jesp.2012.02.008.

Benedetti, F., Carlino, E., & Pollo, A. (2011). How placebos change the patient's brain. *Neuropsychopharmacology Reviews, 36*, 339–354. https://doi.org/10.1038/npp.2010.81.

Bingel, U., Wanigasekera, V., Wiech, K., Mhuircheartaigh, R. N., Lee, M. C., Ploner, M., & Tracey, I. (2011). The effect of treatment expectation on drug efficacy: Imaging the analgesic benefit of the opioid remifentanil. *Science Translational Medicine, 3*(70), 70ra14. https://doi.org/10.1126/scitranslmed.3001244.

Caccialanza, R., Aprile, G., Cereda, E., & Pedrazzoli, P. (2019). Fasting in oncology: A word of caution. *Nature Review Cancer, 19*, 177. https://doi.org/10.1038/s41568-018-0098-0.

Charlier, P., Poupon, J., Huynh-Charlier, I., Saliège, J. F., Favier, D., Keyser, C., & Ludes, B. (2009). A gold elixir of youth in the sixteenth century French court. *British Medical Journal, 339*, b5311. https://doi.org/10.1136/bmj.b5311.

Colloca, L., Sigaudo, M., & Benedetti, F. (2008). The role of learning in nocebo and placebo effects. *Pain, 136*(1–2), 211–218. https://doi.org/10.1016/j.pain.2008.02.006.

Efferth, T., & Kaina, B. (2011). Toxicities by herbal medicines with emphasis to traditional Chinese medicine. *Current Drug Metabolism, 12*(10), 989–996. https://doi.org/10.2174/138920011798062328.

Ernst, E. (2009). Ethics of complementary medicine: Practical issues. *British Journal of General Practice, 59*(564), 517–519. https://doi.org/10.3399/bjgp09X453404.

Ernst, E. (2011). Fatalities after CAM: An overview. *British Journal of General Practice, 61*(587), 404–405. https://doi.org/10.3399/bjgp11X578070.

Ernst, E. (2013). Thirteen follies and fallacies about alternative medicine. *European Molecular Biology Organization, 4*(12), 1025–1026. https://doi.org/10.1038/embor.2013.174.

Ernst, E., & Smith, K. (2018). *More harm than good? The moral maze of complementary and alternative medicine*. Cham: Springer.

FakeMed. (2020). https://fakemedecine.blogspot.com/p/english-version.html

Fraser, B. (2018). China's lust for jaguar fangs imperils big cats. *Nature, 555*, 13–14. https://doi.org/10.1038/d41586-018-02314-5.

HAS. (2017). https://solidarites-sante.gouv.fr/soins-et-maladies/signalement-sante-gouv-fr.

Izzo, A. A., & Ernst, E. (2009). Interactions between herbal medicines and prescribed drugs: An updated systematic review. *Drugs, 69*(13), 1777–1798. https://doi.org/10.2165/11317010-000000000-00000.

Johnson, S. B., Park, H. S., Gross, C. P., & Yu, J. B. (2018). Complementary medicine, refusal of conventional cancer therapy, and survival among patients with curable cancers. *JAMA Oncology, 4*(10), 1375–1381. https://doi.org/10.1001/jamaoncol.2018.2487.

Kruger, J., & Dunning, D. (1999). Unskilled and unaware of it: How difficulties in recognizing one's own incompetence lead to inflated self-assessments. *Journal of Personality and Social Psychology, 77*(6), 1121–1134. https://doi.org/10.1037/0022-3514.77.6.1121.

McIntyre, L. (2019). *The scientific attitude. Defending science from denial, fraud, and pseudoscience*. Cambridge: Massachusetts Institute of Technology.

Mehra, M. R., Desai, S. S., Ruschitzka, F., & Patel, A. N. (2020). Hydroxychloroquine or chloroquine with or without a macrolide for treatment of COVID-19: A multinational registry analysis [published online ahead of print, 2020 May 22] [retracted in: Lancet. 2020 Jun 5]. *Lancet, S0140-6736*(20), 31180–31186. https://doi.org/10.1016/S0140-6736(20)31180-6.

Miguel, M. G., Antunes, M. D., & Faleiro, M. L. (2017). Honey as a complementary medicine. *Integrative Medicine Insights, 12*, 1178633717702869. https://doi.org/10.1177/1178633717702869.

Rief, W., Nestoriuc, Y., Weiss, S., Welzel, E., Barsky, A. J., & Hofmann, S. G. (2009). Meta-analysis of the placebo response in antidepressant trials. *Journal of Affective Disorders, 118*, 1–8. https://doi.org/10.1016/j.jad.2009.01.029.

Chapter 6
Motives and Facilitators of Non-pharmacological Intervention Use

1 Introduction

The use of non-pharmacological interventions (NPIs) has never been so massive around the world. Who are their users? What are their motivations? Do they follow personal intuition, healthcare professional prescription, or media advice? How do they avoid preventive and therapeutic randomness? The choice of an NPI is never completely by chance. Some life events facilitate the decision. This chapter addresses, without ranking by importance, these moments that do not always lead to the best choice.

2 Holistic Need for Health

A mountain race, a trail, a dive, a crossing, a high jump: many occasions to truly challenge yourself are possible in unusual environments. These solitary moments liberated from a fast-paced and crazy occidental life, where you cannot cheat with yourself, are paradoxically rare. The strangeness, the difference, and the contrast arouses curiosity and can open you up to other forms of care that are more genuine, closer to nature, and more in harmony with the body's needs.

A trip, an expedition, or a trek could be an occasion to explore new horizons in terms of health practices and lifestyles. India, the Middle East, Africa, Maghreb, South America, Australia, Asia, and Indonesia have so many traditional practices for humans, highlighted recently by the World Health Organization (WHO) (see below, Adhanom Ghebreyesus 2019). Everywhere humans have sought to heal themselves by all means, with herbal remedies for example (Apelian and Davis

Fig. 6.1 The Dru yoga, an inner NPI. (Published with kind permission of © Plateforme CEPS. All rights reserved)

2018). The discovery of Dru yoga during holidays in India can be an example (Fig. 6.1).

> **WHO Traditional Medicine Strategy 2014–2023**
> *"Traditional and complementary medicine (T&CM) is an important and often underestimated health resource with many applications, especially in the prevention and management of lifestyle-related chronic diseases, and in meeting the health needs of aging populations. Many countries are seeking to expand coverage of essential health services at a time when consumer expectations for care are rising, costs are soaring, and most budgets are either stagnant or being reduced. Given the unique health challenges of the twenty-first century, interest in T&CM is undergoing a revival."*
> Tedros Adhanom Ghebreyesus, Director-General WHO (2019, p. 5).

The inner and holistic exploration of health is motivated by the limits of the reductionistic biomedical approach taken in Western countries. They have limited treatments to a few marketable biomedical products targeted at the diseased organ. Many citizens aspire to be treated differently and especially to fall ill or be dependent less often, whether these disorders are minor or major. Many are disappointed and disillusioned by so many cases of error, unpredictability, fraud, and lawsuits often linked to medical industry pressure or health provider fraud (e.g., incorrect reporting of diagnoses; redundancy of procedures; overutilization of services; unnecessary imaging, testing, and surgery; useless prescription of drugs). For example, the white paper from the U.S. Healthcare Fraud Prevention Partnership (2018) estimated $87 billion in revenue from clinical laboratory services in 2017. While Medicare strictly limits coverage for genetic screening tests, it does cover many genetic tests that meet the criteria for diagnostic tests such as those used to treat cancer and other medical conditions. A high-cost diagnostic test can easily be marketed to the masses. The scammers are using "fear factor" messaging to vulnerable persons anxious to diagnose existing health concerns and proactively identify future medical conditions. This, along with legitimate interest in genetic tests, may help explain the rapid rise in genetic testing fraud. Providers, including physicians, may not even know that they are part of a fraud strategy. A current trend to watch is drug sensitivity testing, known as pharmacogenomics. Pharmacogenomics is the testing of specific genes to gauge how patients will react to certain medications. Healthcare providers are interested in this type of service because it may reduce the risk of adverse drug events. Seniors are often targeted in pharmacogenomic schemes because they may have a number of medical conditions that require several prescription medications. Physicians, anxious to avoid adverse drug events, may order hundreds of pharmacogenomic tests well beyond the scope of the patient's condition. Physicians may also be told by marketers that they can bill for interpreting the test results in order to maximize revenue. Abuses are not only related to biological testing. As of 2016, nearly 198,000 Americans died from overdoses related to prescription opioids since 2000 (Higham and Bernstein 2017). The drug industry and drug distribution companies try to corrupt, and have corrupted, some doctors into illegally prescribing large volumes of pain meds, and some pharmacists into dispensing these pills for a profit. Drug distributors also have facilitated user access to drugs through illegitimate online pharmacies that do not require a prescription. (Higham and Bernstein 2017).

Dissatisfaction
"Medicine has never been so effective. However, never dissatisfaction has been so great among patients and doctors." (Professor Daniel Loisance, Member of the French Academy of Medicine 2019).

Finally, a clientele for NPIs comes from intellectuals and those of a countercultural persuasion attracted from time to time by traditional remedies which have the allure of mystery, symbolism, and ritual. Some NPIs inspired from ancestral and holistic medicine encourage the (re)discovery of the role of introspection (inner self), spirituality (not necessarily religion), and one's life project and relationship with others. These slow, gentle, and less invasive practices, at peace with the rhythm of the body, thus create a space for the release of tensions, a return to basics, often more linked to the proximity of the physical and human environments. Many people want to go further while maintaining permanent contact with modern medicine. They remove barriers that may hinder activating their body's innate healing response. They engage mind, body, spirit, and community to facilitate healing; in short, an intuitive entrance to an *"integrative medicine"* (Rakel and Weil 2018), a *"medicine focused on the person"* (Di Sarsina 2017). Vietnam is an excellent example of a country that brings together its national traditional medicine, which gives a major place to prevention, and a biotechnological curative medicine. An entire hospital is dedicated to traditional medicines in Hanoi. These practices are important sources of inspiration for new NPIs.

> **Advocacy for a "Medicine Focused on the Person"**
> *"The time has come now for a deep reflection involving all the society levels as it is the society itself, the patients' and citizens' associations, as well as the individual citizens (male and female) who demand that the integrity and wholeness of each human being is restored and respected with regards to the diagnostic-therapeutic access."* (Di Sarsina 2017, p. 48).

3   Lack of Resources

There are a number of traditional remedies that the Western-trained doctor ignores, for example, boumezoui, koulchite, or ragued from Morocco. A majority of patients remain rural or semi-urban, due to their education, their sociocultural environment, and their living conditions which distance them from the big cities where doctors are concentrated. Most centers have limited resources to treat them with very expensive modern care. Traditional medicine is one of the primary sources of health care in many developing countries. The availability and/or accessibility of conventional medicine-based health services is, on the whole, limited, compared to that of traditional medicine present on the ground and readily affordable. For millions of people in Africa, native healers remain their health providers (WHO 2013).

Moreover, the use of traditional medicine is adapted to cultural and historical influences that pragmatically answer to health needs. This medicine effectively treats a large number of common troubles (e.g., colic, headache, sprain). This traditional art of healing fits easily into the sociocultural environment and, as a result, it is accepted and therefore socialized (e.g., herbal preparations, ritual practices).

Last, but not least, NPIs are used not only to treat illnesses, but also used widely to prevent diseases, to promote health and to maintain health. These methods have proved to be cost-effective for some governments (Chap. 7). As mentioned by the WHO (2013, p. 16), *"the affordability of most traditional medicines makes them all the more attractive at a time of soaring health-care costs and nearly universal austerity."* With prevailing current global financial constraints, use of NPIs for disease prevention, self-health care, and health promotion can actually reduce escalating biomedical expenditure.

> **Prevention Costs Less than Disease Management and Treatment**
> *"The threat to our societies from chronic, non-communicable diseases cannot be overstated, and the proportion of people affected is growing. Up to 40% of the EU population aged over 15 years report a long standing health problem and two out of three people who have reached retirement age have at least two chronic conditions. Yet these conditions, which are responsible for 86% of deaths in Europe, are largely preventable"* (European Chronic Disease Alliance 2019, p. 17).

4 Testimony

A mode of entry into the use of NPIs can follow reading an autobiography or listening to a successful individual experience. A personal story is so compelling that it leaves indelible traces in the memory. As long as the experiencer does not try to teach a lesson, everyone can identify with her or him. Terry Fox began the run across Canada that he named the "Marathon of Hope." The politician Jean-François Deniau shared his solo crossing of the Atlantic when he was suffering from advanced cancer. The tennis player and singer Yannick Noah popularized fasting. Dr. David Servan-Schreiber (2011), who convinced that he had conquered cancer, has made the general public aware of foods rich in omega 3 and turmeric. Who has not been touched by the videos of Jon Kabat-Zinn on mindfulness? Who has not been passionate about Mihaly Csikszentmihalyi's recipes from positive psychology (e.g., happiness, gratitude, flow)? Who has not been impressed by Jill Bolte Taylor's TED talk when she explains her stroke with a human brain in her hands? Televised talk shows like those have democratized new health practices. This is the strength of the testimony, talking directly to the general public.

The testimony also does good to those who tell of their adventure and their experience. It pushes them to go further, for example, sharing their experience with other patients by making hospital visits, establishing solidarity movements (e.g., cancer survivors), creating associations, etc.

Patients become NPI ambassadors by testifying whenever they can that a physical activity program saved their life, that a psychotherapy made them see things differently, and that a diet transformed them.

> **Terry Fox, the One-Legged Runner, a Hero across Canada and around the World**
>
> Terry Fox was a Canadian humanitarian athlete who had his leg amputated 15 centimeters above the knee in 1977 following osteosarcoma, a type of bone cancer. He said "nobody is ever going to call me a quitter" (Scrivener 2000). Equipped with an artificial leg, he decided to run his "Marathon of Hope," a run across Canada to raise funds for cancer research. He started on April 12, 1980 in Saint-Jean de Terre-Neuve and covered a distance of 5373 km for 143 days (nearly 42 km per day) before a metastatic recurrence of his cancer affected his lungs. He has become an icon for all cancer patients, far beyond Canada. His marathon of hope continues to live in the four corners of the globe each year, well beyond his death.

Successful experiences with NPIs can generate expertise to help other patients and healthcare professionals. To illustrate, a remarkable initiative founded by Catherine Tourette-Turgis (2020) is the "University of Patients" at the University of Paris, Sorbonne. This innovative training program consists of integrating "patient-experts" from the world of associations and organizations into university degree courses in therapeutic education. Its purpose is to recognize the experience and expertise of patients. The diploma facilitates (1) the participation of "patient-experts" in return-to-work strategies, (2) mutual assistance procedures, (3) the integration of patients into the training of health professionals, and (4) patient participation in the organization of care.

Patients organize rallies. They get involved in charity events. They participate in conferences to report on the benefits of certain NPIs. These movements come together in associations, in networks, and in supranational federations. For example, a pilot study of an inpatient pulmonary rehabilitation intervention indicated that a combination of services coordinated within self-help associations federated in a collaborative regional healthcare network called AIR+R, which was successful in providing an innovative, efficient, and cost-effective maintenance program, not only for maintaining but also for improving the benefits of such a rehabilitation (Moullec et al. 2008). Each local multidisciplinary team provided similar coordinated sessions. To ensure the quality and the consistency of the maintenance program, all professionals participated in training on pulmonary rehabilitation, during which they were taught all components of the program as follows: (1) individualized exercise training (3.5 h/week) supervised by an exercise teacher in a gymnasium including breathing exercises, interval training (in circuit and in team sport, in line with the patient's interest), strength training, upper limb training (e.g., weights, elastic bands), and endurance training with nature walking at the ventilatory threshold; (2) health education provided alternatively by all professionals of the healthcare network (2 h/month) in a municipal conference room; and (3) psychosocial support (with discussion group 1 h/month) supervised by a psychologist in the same room. The self-help associations could obviously decide to add discretionary leisure activities desired by members, such as museum visits, restaurant outings, or inter-group meetings. These groups were entirely run by the patients. They benefit from resources provided by the regional network financing the 96 sessions supervised by the health professionals in the 1-year maintenance program. The complementary and coordinated services and the collaboration of practitioners, including the lung specialist, contributed to a coherent strategy for efficiently responding to the chronic nature of the chronic obstructive pulmonary disease, its systemic manifestations, and the unpredictability of symptoms. Monthly educational sessions associated with weekly exercise training enhanced practical knowledge and behavior. The correspondence between the technical language of medicine and what the patients felt and imagined about their disease was even strengthened. The resulting reduction of semantic discrepancy facilitates the long-term maintenance of healthy behaviors and could be a potent source of motivation for patients to become active and responsible participants in their own care. The knowledge of regularity and the enhancement of motivation in health behaviors may explain the continuing improvement in exercise performance and quality of life over the entire post-rehabilitation year. In addition, a monthly psychosocial group session led by a clinical psychologist would be an opportunity for the patients to integrate the functional progress made during exercise sessions and to improve overall well-being and self-esteem. Moreover, belonging to a healthcare network with a self-help association contributed to the feeling of being included in a community. The community is likely to answer to psychological isolation with coordinated activities and collective projects, thereby decreasing depressive, morbid thoughts. The social support of other participants and the greater partnership with health professionals might encourage individual initiative and attempts to reduce emotional reactions and choose adapted behaviors. The patients viewed themselves as active partners in the treatment process (Moullec et al. 2008).

5 Artistic Inspiration

Who has not been touched by sketches by Anne-Alexandrine, known as double-A, suffering from multiple sclerosis; by Jeremy Demay, who has overcome depression; or by Djamel Debouze evoking his infirmity? And who has not been "hypnotized" by a Messmer show? Ludwig van Beethoven was deaf. Becoming deaf before his 30th birthday, the German composer wrote his famous ninth symphony when he was already deaf! All his life, Beethoven tried to hide his deafness while continuing to compose his works. At the end of his life, he preferred to hide at home rather than reveal his handicap to his loved ones.

These authors and artists succeed in transmitting medical information related to their personal story. These shared experiences, so authentic and so intimate, are a source of inspiration, a source of hope and resilience suggesting that art can be a research field for therapy. They push patients to try art therapy interventions such as an expressive writing intervention, a music therapy program, or drama therapy. For example, the smartphone-based application called Music Care (2020) offers individualized music intervention sessions using a standardized protocol following international scientific guidelines. The program allows patients to listen to a variety of professionally recorded music pieces of their choice. The app has previously been evaluated and showed significant effects in reducing pain and anxiety (Guétin et al. 2012). A randomized controlled trial reported benefits on physiological variables, such as heart rate, breathing, blood pressure, and bispectral index (Guétin et al. 2012). In the context of chronic pain, the results of a controlled study showed promising results on reducing the intensity of pain and specifically on decreasing the consumption of anxiolytics and antidepressants (Guétin et al. 2009). In the music therapy group, the proportion of patients consuming anxiolytics decreased from 90.9% to 42.9% in a 60-day follow-up (−53%). In the control group these proportions are respectively 79.1% and 66.7% (−16%).

Writing, Painting, Illustrating, Singing, Music Playing, Dancing: Inspirational Sources for NPIs

Jean-Jacques Rousseau, the eighteenth-century French writer, complained of many ills that are difficult to pinpoint. Several doctors have hypothesized acute intermittent porphyria, a disease affecting the central nervous system (a rare disease with pain in the arms and legs, vomiting, confusion, constipation, tachycardia, fluctuating blood pressure, urinary retention, psychosis, hallucinations, and seizures).. The philosopher wrote about his illness: *"I was born crippled and sick, I cost my mother her life and my birth was the first of my misfortunes."*

Edouard Manet, the nineteenth-century French painter, had dazzling pains in the legs that became more and more frequent. The crises prevented him not only from walking but also from standing and therefore from painting.

Vincent Van Gogh, the nineteenth-century Dutch painter, suffered from schizophrenia and bipolarity. He cut off his left ear during a seizure and committed suicide seven months later. He is one of the greatest painters.

Henri de Toulouse Lautrec, the twentieth-century French illustrator, suffered from pycnodysostosis, a disease that limits growth and makes bones fragile. He devoted his life to describing Parisian life at the time, and in particular *"the soul of Montmartre."*

Frida Kahlo, the twentieth-century Mexican painter, had suffered from poliomyelitis since the age of six. Made worse from a disastrous road crash, her health forced her to lie in bed or in a wheelchair. To overcome boredom and pain, she painted works on bodily themes such as abortion, sexuality, injured flesh, and suffering.

Glenn Gould, the twentieth-century Canadian composer and pianist, had autism (Asperger's syndrome). He produced unforgettable musical works.

Ray Charles, the twentieth-century American singer, songwriter, and pianist, had been completely blind since the age of 7. Despite this disability, he became an international star of jazz, gospel, blues, and country. He is considered one of the pioneers of soul.

Alicia Alonso, the twentieth-century Cuban dancer and choreographer, became partially blind from the age of 19. She persevered in dancing. Thanks to the lights, she managed to orient herself, and to work with her partners so that they would be exactly where she would think they would be during the performances. She directed the National Ballet of Cuba.

6 Human Performance

There are so many people on this planet who are seeking to push their limits, driven by a model of society based on competition and profitability. It is, therefore, not surprising that these people are major consumers of NPI: athletes, e-sports performers, artists, business leaders, soldiers, professionals, or amateurs. They use boosting products in massive doses, drastic diets, and extreme training methods. The only thing that counts is the performance, the entertainment, the differentiation. What counts is the feat being listed in a world record book, from the most venerable to the most pathetic, each with its own scale of values. Some humans conquer vertiginous mountain faces, some others accumulate medals, while some others prefer to throw cherry stones with their mouth. It is common to see athletes with two bags, one filled with classic items, the other filled with food supplements, neurostimulation devices, a hypoxic chamber, rubber bands, instruments for biofeedback, and a personal pillow. In a bodybuilding gym or fitness center, it is common to find spaces selling products (e.g., protein powder, food supplements, anti-fat capsules, boosters, energy drinks, herbal remedies, and detox cure books).

These avid pushers are ideal beta-testers for NPIs. They assess, combine, personalize, and optimize NPIs by trial and error. Regardless of the consequences for health and personal life, they take full responsibility for it in case of failure. No matter what the sacrifices, they follow through on their ideas, at whatever the cost. Their failure is ultimately just a step before the success. They never give up, start again, bounce back, and find new resources to go further in their quest for the absolute, their path to glory. They find new technical solutions, new dosages, new formulations, new combinations. They are individuals with the excuse that it only works for them. They try to go beyond the possible. Their challenges are inspired by the Olympic motto endeared to Pierre de Coubertin, *citius, altius, fortius* (faster, higher, stronger).

As if that were not enough, performers often live with people encouraging them (e.g., coach, manager), strengthening them (e.g., doctors, psychological trainers, physical trainers, dieticians, material engineers), admiring them (e.g., fans, followers), stimulating them (e.g., family, agents, adversaries, humanitarian foundations), writing them down in history (e.g., journalists, former practitioners, historians, federations, politicians), and financing them (e.g., companies, associations, clubs, sponsors). Doctors trained in the treatment of diseases (classic restorative medicine) are passionate about positive medicine, a reverse medicine potentiating human capacities, both structures and functions. Many doping products come from the diversion of an initial therapeutic use such as erythropoietin, for example. Diets, isotonic sports drinks, high-energy sports drinks, electrolyte supplements, sports gel, protein supplements, sports bars, sports confectionary, liquid meal supplements, plants, psychological methods, physical techniques, digital tools, everything goes into the hands of these compliant beta-testers. Suffice it to say that manufacturers are delighted to have guinea pigs, ready for anything, to try their latest innovation. They are hardworking, available, pragmatic, reactive, and ready to report on their experience; demanding, rigorous, sensitive, willing to test an innovation for free, resistant to side effects, and discreet. What more could you wish for?

In search of the absolute, or some would say in search of the absurd, the failures of some individuals serve the performance of others. It does not matter if these performances hide the reality of many hours of preparation. Empiricism and pragmatism are in order. Good practices are transmitted orally, generation after generation. This trend is growing; globalization and the opening of communication systems do the rest. Anti-doping control agencies will always lag behind human ingenuity.

Sports Foods and Supplements

Numerous nutritional products are marketed with claims of optimizing athlete health and function and/or enhancing performance. Products that fall under the banner of "Sports Foods" or "Dietary Supplements" may be used to support performance during training and competition or for enhancing aspects of training adaptation, recovery, immune function, and/or overall athlete health (Peeling et al. 2019). Effective marketing campaigns and athlete endorsements may convince us that certain sports foods and supplements are fundamental in allowing athletes to reach their sporting goals:

- Weight management.
- Nutritional ergogenic aids.
- Dietetic agents favoring muscle mass accretion.
- Nutritional protocols that favor mitochondrial biogenesis and functioning.
- Nutritional means to increase muscle and/or blood buffer capacity.
- Dietetic strategies for allergy, intolerance, digestive system disorder.
- Nutritional methods for improving recovery, injury repair, and return to play.
- Dietetic protocols for improving sleep.

Finally, these performers in search of perfection are only responding with all means to a society requiring the extraordinary of ordinary people, admiring incredible talent (e.g., TV shows), and sanctifying *Homo deus* instead of the simple *Homo sapiens* (Harari 2017).

7   Love

Falling in love is a nice way to encourage you to take care of yourself to please your loved one. This very special state, bathed in hormones, to which everyone aspires, is an excellent moment to encounter various NPIs. So, yes, there is often an aesthetic ambition at first: to refine their silhouette, modify an unsightly area, firm and tone the muscles, heal their skin, improve gait, and straighten their posture.

While the period of euphoria passes, NPIs can settle permanently in the lives of couples to maintain the flame and address the misunderstandings between couples that are so well described in some bestsellers like the 1992 book *Men Are from Mars, Women Are from Venus* authored by the American relationship counselor John Gray.

However, love is not always eternal. Separation is, therefore, a difficult ordeal to go through. Wanted or not, it leaves indelible marks and voids. It restructures relationships with children and the family. It upsets relations with friends. It opens up trying new NPIs such as specific massages, phytotherapy protocols, and detox sessions.

8   Parenthood

The arrival of new life is a special and unique occasion. Most parents talk about it as the greatest moment of their lives, with no going back. This is an opportunity to take care of yourself in addition to the baby. It is also a time to be part of a family lineage, to transmit value, and to take responsibility. It is a great opportunity to change health behaviors; for example, with regard to tobacco use. The "hormonal bath" in which mother is immersed during pregnancy is a great opportunity to quit smoking. A specialized NPI such as minimal counseling during a 15-minute session administered by a healthcare professional reduces the smoking rate of expectant mothers by 30% (Melvin et al. 2000). Why not systematize it in maternity hospitals?

NPIs increasingly play a role during key life transitions such as transition between childhood and adolescence, transition between adolescence and adulthood, and the parenthood transition (Young and Amarasinghe 2010). For example, a parent-led NPI combining psychosocial techniques with novel therapeutic elements based on developmental models of social and cognitive development treats the core symptoms of attention deficit hyperactivity disorder and associated oppositional and non-compliant behavior (Thompson et al. 2009).

9   Professional Failure or Success

Professional success or failure can trigger emotions that dramatically alter the person's life course. These emotions, particularly in overworked urban residents, arouse desires for elsewhere, or at least a balance sheet of competence, health, heritage, and succession. They mark the passage of time and encourage people to change their lives before it is too late.

> **An Insight for the Rest of Life**
> In these moments described above and below, for too many people the encounter with NPIs is inevitable. These moments reveal vocations and desires that are muted in the course of a daily life flooded with ready-made solutions and emergencies where this is not the case.

Professional successes or failures make the way for the human absurdities and contradictions so well described by Jean D'Ormesson (2019). Companies offer book summaries in 20 minutes. Podcasts invite you to listen to your favorite radio shows at high speed. Buying platforms encourage you to buy your trip or concert two years in advance.

Fortunately, NPIs are putting quality and time back at the center of human life. They reveal the small signs launched by the body but which remained without echo. It is finally a chance to get rid of these evils of burnout, overdose, depression, and related psychosomatic words that ruin life (e.g., back pain, intestinal transit problems, tinnitus, headache, rashes, sexual breakdowns).

10   Early Screening

The therapeutic progress of medicine is due in particular to early detection. Many diseases are detected years before the signs are observable by a healthcare professional and/or felt by the patient, thanks to biological or neurocognitive tests. This is true for common chronic diseases such as diabetes, neurodegenerative disease (Prince et al. 2011), heart disease, and respiratory disease. This also applies to rare diseases; for example, in children in France under the impetus, in particular, of the Association for the Screening and Prevention of Disabilities in Children. Testing is primary for cancer where personalized therapies are conditional on not only the presence of a mutated gene of interest in the tumor but also at the constitutional level (Pujol et al. 2018). In addition, through machine learning (in the field of artificial intelligence), it will be possible to extract patterns within patient data and exploit these patterns to predict patient outcomes for improved clinical management.

Predictive medicine leads to treating the disease before the patient is suffering (asymptomatic). Even if it may come as a shock when the result is announced (especially if the disease is discovered by chance), this precocity becomes a chance for the patient to live better and longer. It creates a chance to really listen to the body, to take a break, and to know the inner self.

> **The Example of Angelina Jolie**
> The actress wrote a courageous article in *The New York Times* to raise the awareness of women who may have the BRCA1 or BRCA2 gene. These mutations increase the risk of breast cancer before the age of 50 by 72% for BRCA1 and 69% for BRCA2 (Kuchenbaecker et al. 2017). She decided in 2013 to have her breasts removed after a blood test that revealed a BRCA1 genetic mutation. Two years later, in agreement with her doctors, she underwent additional surgery to have her ovaries and fallopian tubes removed.

The announcement of a disease following early detection is a moment of realization that life is short and that it must be prolonged as long as possible. This period can make the person think of time and relationships with others differently; it promotes entry into NPIs that could prevent, or delay, the course of the disease. Early diagnosis is a chance for a new or second life.

11   A Diagnosis or Major Risk

The announcement of an illness is always a shock. It creates a before and an after. It forces us to think and to rethink its trajectory, to reconsider the future.

It opens the door to the use of NPI, by the time it gives and the relativity it brings to things that were considered so important in the past and that now have become so paltry.

It means we question life and death, the useful and the futile, making psychotherapy more accessible, for example. It changes the way of life. The French philosopher Michel Onfray joked for a long time that his only physical activity was going down to the cellar to get a bottle of wine. The occurrence of a stroke led him to reconsider the place of physical activity in his daily life.

The COVID-19 episode in 2020 is having a lasting impact on Western countries, including making people better consider their health. Asian and African countries have already been sensitized to previous aggressive viruses (e.g., H5N1, SARS). Large-scale quarantine has made the population, as a whole, understand how there was a collective responsibility regarding the propagation of the virus with the use of relevant NPIs (e.g., certified protective masks, hand-washing procedures, social distancing protocols) and the maintenance of their own health (e.g., physical activity program, diet).

12   Comprehensive Relationship with a Practitioner

The world is experiencing such a general social acceleration that even the general medical consultation time has been reduced to 12 minutes on average. This time seems so constrained that the patient and the doctor go directly to the concrete and surface facts, leaving aside the intimate, the deeper issues, the doubts, the beliefs, the real motives. The patient perceives the doctor as a prescription administrator, and no longer as a caregiver who approaches the person as a whole. Some experiences in oncology have become caricatural in some cancer centers. The surgeon or the oncologist is focused on the tumor without thinking about the functional, psychological, and short- and medium-term impacts, without enough discussion during the testing, announcement, treatment, post-treatment, and control phases. The patient accuses the doctor of forgetting the paramount psychosocial dimension, largely mentioned in the WHO definition of health.

> **The Forgotten Relationship-Centered Care**
> *"It is much more important to know what sort of a patient has a disease than what sort of a disease a patient has."* (William Osler 1849–1919).

On the contrary, NPI practitioners have more time dedicated to their patients than biotechnical professionals assimilated to procedural engineers and impersonalized robots. They spend time addressing psychosocial issues, leaving patients more satisfied. This compassionate and empathic approach emphasizes health enhancement in the treatment of the disease and being proactive in addressing early warning signs and risky lifestyle factors. The emphasis is on health promotion as an integral part of disease treatment. Most NPI professionals contribute collaboratively to health enhancement with attention to stress management, spirituality, issues of meaning, nutritional habits, exercise, fitness, addiction (e.g., tobacco, cannabis, alcohol use, cocaine, other drugs). The close and confident relationship is particularly effective for patients with chronic disease (Chap. 3). Many patients find that the more they incorporate NPIs into their lives, the less difficulty they have in managing chronic disease no matter what the cultural orientation. Patients are strongly motivated to impact their disease, health, and life. They believe strongly, sometimes incorrectly, that changes are possible.

> **Confidence or Credulity?**
> Privacy, availability, and comprehensiveness are at the heart of the training of "alternative medicine" practitioners (Health Canada 2001). They are all the more a contrast to today's technological and sanitized medicine. But, to "take care" of health should not be done anyhow, anywhere, anytime, and with anyone. It is important to be alert to some advice given by self-taught people, always rehashing the same successful experiences and skillfully spacing out visits to the attending physician.

13 Hospitalization

A hospitalization creates a break with everyday life. It never comes at the right time. The period is worrisome, stretching the time for reflection, about health, well-being, relationships, life.

> **An Increased Prescription of NPIs**
> More than 30% of doctors working in hospitals are interested in NPIs. They are prescribing more and more NPIs to relieve pain (e.g., targeted hypnosis protocol), to improve sleep (e.g., food supplement with lime, lemon balm and verbena diluted in a glass of water at bedtime), to facilitate intestinal transit (e.g., manual therapies), to reduce stress (e.g., cardiac coherence methods), to calm burns (e.g., natural ointment), to prevent muscle wasting (e.g., a specific online muscle reinforcement program), to facilitate the elimination of chemical and/or toxic products (e.g., green tea), to improve knowledge on the management of a chronic disease (e.g., serious games), or to facilitate the relationship with caregivers (e.g., music therapy).

Studies are looking at traditional Chinese massages to reduce restlessness and anxiety in Alzheimer's patients or therapeutic Qigong to relieve pain caused by chemotherapy. Manual therapists relieve the pain of burns from radiation therapy. A professor of medicine at the Salpêtrière Hospital in Paris, Alain Baumelou, has developed an acupuncture practice derived from traditional Chinese medicine to reduce anxiety before surgery.

Once integrated, these practices persist even in the event of a change of department head or hospital management. Patients are asking for more. Such NPIs leave a positive imprint on patients and their families.

14 Biomedical Treatment Failure and Side Effects

Many people have become skeptical about, and distrustful of, medical organizations and the biomedical world industries. The lawsuits against pharma and medical equipment and biological testing companies have justified their disapproval. They preach for "natural remedies", perceived, sometimes incorrectly, to be without risk. Some want to avoid drug consumption or vaccination. Some are looking for a solution to decrease biomedical treatments' side effects, such as a response of acute (e.g., anxiety, headache, nausea, depression episode) or chronic (e.g., pain, fatigue, stress, tension) symptoms. Some want to compensate residual disabilities after biomedical treatments (e.g., sequels, esthetic disfigurement). Some want to try another path after a treatment failure (e.g., cancer recurrence, palliative care, surgery failure). Some want to try an innovative strategy because no biomedical treatments exist (e.g., rare diseases). Most NPI users look for a way to achieve better living, increased comfort, social support, and hope for the future. Most of them interpret WHO's basic definition of health as well-being. In addition, most users want to make up for the lack of overall care (physical, psychological, even spiritual) and time spent listening to caregivers.

Beyond therapy, the doctor is sometimes perceived as very technical, more prescriber than caregiver, sometimes losing sight of care as a whole process (especially its social dimension) and with empathy (e.g., compassion, listening, availability).

To be clear, the vast majority do not refuse conventional medicines. In oncology, for example, patients continue to trust their oncologist and their treatment, but they want additional help from NPIs. This is typically the case for cancer (Lognos et al. 2019).

15 Pain(s)

In 80% of cases, the reason for consulting a general practitioner is acute or chronic pain. *Pain is always a question, and pleasure is the answer.* Pain encourages us to look for all possible ways to soothe it.

About 80% of the western population develop low back pain at least once during their lifetime. Pain drugs are usually associated with a variety of adverse side effects, such as constipation, delirium, hyperalgesia, myoclonus, nausea, opioid dependency, respiratory depression, sedation, sexual dysfunction, and urinary retention. Pain has a complex origin, with a high variability between individuals and over time. The chronicity and uncertainty of the manifestations are a major risk. It requires multidisciplinary and multifactorial support. Mild-to-moderate pain may be relieved by NPIs alone.

> **Important**
> For a decade, medical practice has opened up to other paths than chemistry; for example, acupuncture, osteopathy, medical hypnosis, mindfulness meditation, and music therapy.

Teams specializing in the assessment and treatment of pain have been set up in university hospitals, such as the one led by the doctor Patrick Ginies in Montpellier (France). In this case patients discover NPIs (e.g., yoga, neuromuscular electrical stimulation, heat/cold application, cognitive-behavioral method of therapy), and their active mechanisms (Chap. 4). Once the pain has been relieved, they use other NPIs for prevention.

16 Fifties and Healthy Aging

Middle age is a key period for health (Myint et al. 2011). A U.S. study showed that a change in these five behavioral factors (never smoke, have a normal weight, do physical activity regularly, have a healthy diet, consume alcohol in moderation) prolongs life expectancy by 14 years for women and 12.2 years for men, respectively, by the age of 50 years (Li et al. 2018).

Fifty years is an age when anything is possible, an age when diseases are not present and can be avoided or at least delayed by several years (Steptoe et al. 2015). It is an age when we can act against the pressures chosen (family, work, leisure, etc.) or suffered (retirement, politics, etc.). It is an age when we can cut back on certain foods (high in fat, sugar or sodium), focusing on simple ingredients and fewer processed foods, avoiding artificial ingredients, hormones or antibiotics, genetically modified organisms, and chemical products.

Too many prevention programs (memory workshops or fall prevention, for example) are offered too late to the elderly. The expected effects are not obtained because the injuries or handicaps are irreversible.

Researchers are working to develop new integrative outcomes when the disease is not yet there, but where the physiological and psychological resources to adapt to its environment are starting to run out. It is about the notion of frailty (Fried et al. 2001), for example, to prevent cognitive decline (Yao et al. 2020).

> **Frailty**
> It corresponds to a state of vulnerability, to secondary stress, to multiple deficiencies in the body systems that lead to a decrease in physiological and psychological reserves. Strengthening your reserves allows you to go further: for memory, concentration, stress management, physical capacities, flexibility, balance, good humor, and general well-being.

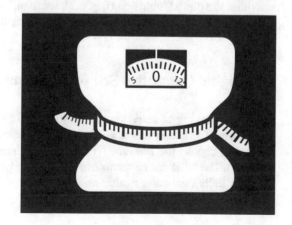

Fig. 6.2 Fasting: An increasing field of interest for healthy aging and care. (Published with kind permission of © Plateforme CEPS. All rights reserved)

Fig. 6.3 Digital tools: Real-time data monitoring systems for better implementation and dissemination of, and adherence to, NPIs. (Published with kind permission of © Plateforme CEPS. All rights reserved)

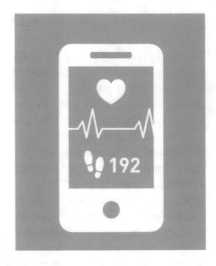

NPIs are specifically aimed at 50-year-olds, fragile or not. The Japanese population counted no less than 67,824 centenarians in 2017! A record. People in Okinawa, for example, live longer than almost anyone on earth. In a group of islands known as the "land of the immortals," called Okinawa, two-thirds of residents who reached 100 years of age were still living independently at the age of 97. Okinawans have low rates of heart disease, stroke, cancer, and diabetes.

Having an overall good experience with NPIs supports the fashionable concept of "anti-aging medicines". They are all the rage in shops and specialized centers to answer to the demands (Stuckelberger 2008). And this is only the beginning. The generation over 50 is asking for NPIs because it is aging radically differently from previous generations with increased longevity and substantial purchasing power. The number of people over 60 is expected to reach 2 billion worldwide by 2050, or one in five according to the WHO (2017). This generation of baby boomers favors positive health, well-being, and nutrition. This demand is at the heart of the *silver economy*. If hospitals and city medicine are gradually integrating NPIs into their activities, the digital, nutrition (Fig. 6.2), tourism, mass distribution, insurance, and accommodation industries for the elderly are spreading them widely, at speed.

17   Digital Quantified Self for Self-Empowerment

Real-time tracking of behavior, perceptions, and biological indicators can be done at low cost. The health data collected makes it possible to better understand health needs and choose the relevant NPIs for the context and lifestyle of each person (Fig. 6.3). The NPIs can be combined and adjusted according to feedback, each having their individual care, health, and life journey. The quantified self is deployed widely nowadays and will increase the adoption of 4P medicine including NPIs (Chap. 2).

While the term "digital therapeutics" may sound futuristic, these tools are already a reality. Digital therapeutic products as NPIs (e.g., serious games, virtual therapies, disease management applications, digital pill boxes) or NPI delivery systems (e.g., case management, healthcare videoconference), are integrated across the healthcare ecosystem. Digital therapeutic products will increasingly influence the way health care is delivered and consumed across the world. For patients, healthcare providers, and payers, it will be important for them to better understand and have confidence in these groundbreaking products. There are a growing number of digital therapeutics on the market today that are being developed in accordance with internationally recognized design, quality, and manufacturing standards. The products aim to directly address patient needs, are safe and effective, demonstrate positive clinical and health economic outcomes, and influence the delivery of health care in a meaningful way (DTX Alliance 2018).

Patients Are Not Stupid
In cancer, for example, rare are the patients who consult an unconventional therapist before a physician and an oncologist (Dilhuydy 2003). Only 4% of Europeans exclusively use alternative medicines, while more than two-thirds use them as supplements to better support treatments and increase their chance of recovery (Molassiotis et al. 2005).

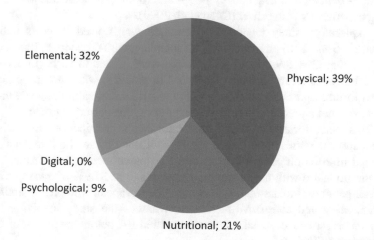

Fig. 6.4 Categories of NPI-related terms mentioned by French women on breast cancer-specific social networks (264,249 posts between 2006 and 2015). (Lognos et al. 2019)

18   Social Networks

Facebook was created in 2004. Since their availability, how much time do you spend on social media every day? Health is a major topic for social media and networks, especially weakly regulated methods such as NPIs. Patients share their personal experiences to receptive communities. The digital social platforms extend NPI practices such as plants and mushrooms from faraway countries, electronic commerce in food supplements, and innovative solutions by way of startups. Patients claim that their failing to mention NPIs to their medical specialist or general practitioner stems from their providers' lack of questioning on the topic, their lack of interest, an anticipation of their disapproval, or their presumed inability to help in this way. With respect to cancer, for example, patients mentioned several reasons for not communicating NPI use: the lack of information given on NPIs for the management of cancer (61% of cases), the lack of questioning from their oncologist (60%), the thought that this does not concern the doctor (31%), the possible lack of understanding of their use by their doctor (20%), the perceived risk of probable disapprobation of their doctor (14%), and the risk that their doctor would no longer take care of them (2%) (Eisenberg et al. 2001). These failures of communication regarding the use of NPIs carry risks during and after cancer treatment, with NPIs potentially generating adverse effects, deleterious interactions, and noncompliance with treatments (Awortwe et al. 2018).

In 2018, three billion people used a social network; this is about 40% of the world's population. Twenty percent of discussions on these networks are related to health. Patients find a space for open dialogue among peers. These platforms also allow an exchange and appropriation of medical information; in particular, to seek answers when they have not been provided by health professionals. This need is particularly pervasive in patients with cancer treated with complex and combined therapies. About 35% of health focus groups and forums are dedicated to cancer and sharing experiences with cancer (Chou et al. 2009).

An exploratory study identified NPI-related words used in posts published between 2006 and 2015 on a French health social network (Facebook, Breast cancer forum). Among 264,249 posts, 27,279 words were related to NPIs (Lognos et al. 2019). NPIs were clearly a topic of concern in patients, with similar proportions between forums and Facebook sources (Fig. 6.4). Digital solutions were the only NPI that was not a preoccupation of women during this period in France.

Patients seek information about the best choices and practices of NPIs with many different sub-categories of NPIs. The results confirmed those of a European survey conducted in 2005 on complementary and alternative medicine by using a structured questionnaire with use reported by an average of 35.9% of a sample of 956 European patients (Molassiotis et al. 2005). The European study had identified 33 complementary and alternative medicines, whereas the study involving French women on breast-cancer social networks identified 101 belonging to four of the five NPI categories (Table 6.1).

Table 6.1 Example of NPI-related terms mentioned by French women on breast cancer–specific social networks

NPI Category	NPI Subcategory	Example NPI-related terms
Physical	Exercise programs	Shiatsu, yoga, tai chi, body building, Pilates, hatha yoga, Iyengar yoga
	Physiotherapies	Speech therapy
	Manual therapies	Acupuncture, acupressing, osteopathy, reflexology, auriculotherapy, chiropractic
Nutritional	Dietary supplementations	Alpha linolenic acid, iron, gamma linolenic acid, amino acids, magnesium, minerals, niacin, ascorbic acid, palmitic acid, creatine, fish oil, biotin, calcium, bioflavin, vitamin (A, C, B, B1, B2, B3, B6, B12, D, D3, and E), multivitamin, folic acid
	Nutritional programs	Dukan diet, fasting, micronutrition
Psychological	Art therapies	Musicotherapy
	Psychotherapies	Hypnosis, hypnotherapy, self-hypnosis, autosuggestion, sophrology, support group, mindfulness-based stress reduction
	Health education programs	Tobacco cessation
Elemental	Cosmeceuticals	Wig, makeup
	Electromagneticals	Chromotherapy, light therapy, quantum medicine, electrotherapy, magnets
	Botanicals	Aloe vera, aromatherapy, belladonna, calendula, chamomile, cinnamon, milk thistle, clove, echinacea, eucalyptus, feverfew, devil's claws, mistletoe, herbs, hops, linseed oils, essential oils, hypericum, kava, lavender, alfalfa, marijuana, peppermint, St. John's Wort, blueberry, passion-flower, dandelion, elderberry, teas, red clover, valerian, cranberry, pomegranate, bitter orange, wild yam, grapefruit, cocoa, noni
	Minerals	Stone

Source: Adapted from Lognos, B., Carbonnel, F., Boulze-Launay, I., Bringay, S., Guerdoux-Ninot, E., Mollevi, C., Senesse, P., Ninot, G. (2019). Complementary and Alternative Medicine in Patients with Breast Cancer: An Exploratory Study of Social Network Forums Data. Journal of Medical Internet Research Cancer, 5(2), e12536. Tables 2–5. doi: https://doi.org/10.2196/12536, licensed under the terms of the Creative Commons Attribution License (https://creativecommons.org/licenses/by/4.0/). ©Béatrice Lognos, François Carbonnel, Isabelle Boulze Launay, Sandra Bringay, Estelle Guerdoux-Ninot, Caroline Mollevi, Pierre Senesse, Gregory Ninot. Originally published in JMIR Cancer (http://cancer.jmir.org), 27.11.2019

The study confirms the ability of social networks to address more deeply and broadly the NPI spectrum compared with a questionnaire survey. Indeed, the questionnaire may hinder patients from revealing their real uses and/or concerns about practices decried by some health authorities (e.g., cannabis). The analysis of posts makes it possible to point out original practices from massive data. If a social network reinforces personal convictions, it offers patients the opportunity to discover new practices consistent with their own beliefs.

19   Palliative Care

In all countries, it is curious to note that NPIs are better accepted by medical professionals and the health authorities at the end of life, compared to primary prevention or classical care. At the end of life, when there is nothing to lose, the place of the "human caregiver" takes back its rightful place. Original and personal methods can be tried to reduce pain, to ease tensions, to reduce fear, to encourage forgiving certain loved ones, to increase exchange, and to provide opportunities for patients to express themselves, to give meaning, and to say goodbye in peace.

It is then an opportunity for exceptional relationships with medical doctors, caregivers, nurses, physiotherapists, manual therapists, aromatherapists, psychologists, dieticians, social workers, etc. These medical, paramedical, psychological, social, and spiritual professionals, in these intense moments, demonstrate extraordinary human qualities and use more and more NPIs (e.g., massages, essential oils, psychological methods, herbal teas, art therapy). Many qualitative studies underline how many patients with advanced cancer claim to benefit from physiotherapy programs (Mas et al. 2015).

They will, of course, help patients, in the noble sense of the term, to better live their end of life and leave in peace. They also enlighten families on the power of NPIs.

20   Death of a Close Person

Last, but not least, the death of a spouse, a loved one, or a close friend is a tragedy. This demise is all the more violent if it is caused by a sudden accident, a surprise attack, or an unexpected suicide. Beyond the lack, the sudden character of death raises questions about the reasons, about our role, about our shortcomings, about the meaning of life and, finally, about our own death.

> **Lifestyle and Early Death**
> Diets containing too many calories are common in Europe and North America, where fast foods and soft drinks are increasingly popular. In countries such as Italy and Greece, where a traditional diet containing large amounts of fruit and vegetables has been replaced by foods high in carbohydrates and saturated fats, obesity, diabetes, cancer, and coronary heart disease are on the increase.

Such an event provides an opportunity to change one's life and to turn to NPIs. Such interventions may enable a return to one's self, meaning one can better face a life that pushes us beyond a stage in our lives that we no longer live.

21 Conclusion

Motives to use NPIs are massively heterogeneous within and between countries: from those who distrust the medical system and the biomedical industry and consume natural products perceived to be safe, to those who want to live in peace, to improve their appearance and treat themselves differently; from those who refuse to age like the young seniors of Generation X and grandpa "boomers," to entrepreneurs of Generations Y and Z who want to push their limits while flourishing on a daily basis; from those who refuse to get sick every month, to geeks of the quantified self who frantically record and follow their biological constants; from those who are disappointed in biotechnological medicine because of various events such as treatment failure, lack of response, or chronic pain, to those who still believe in it, provided that they combine traditional medicine with supportive care (e.g., better living, comfort, decreased side effects, maximization of recovery chances). Whatever their motivation, a large majority of individuals turn to NPIs in addition to authorized biological treatments. And this is made with discernment when the disease is serious, contrary to what is disseminated in some media referring to the motivation for use of NPIs.

Most users of NPIs feel satisfied, although some recognize that some methods do not always work and that they sometimes had to change therapists before finding the one who satisfied them. The feeling of relaxation and well-being is as important as the therapeutic result itself (Centre Fédéral d'Expertise des Soins de Santé 2011).

This rising interest in NPIs reflects not only the demand for holistic, personalized, and preventive health care, but also the compulsory transformation of twenty-first century society, as it is more globalized, more responsible, more populated, and more depleted of resources. This evolution changes the understanding of (1) the body conceived as a whole in interaction with the physical and social environment, (2) the diversity of health preferences due to "boomers," (3) relationship-centered care as a priority, and (4) the advent of "personalized wellness" instead of standardized care. Social networks and quantified-self data will massively influence citizens in this direction.

Key Points

The increase in demand for NPIs reflects a motivation of all generations to be treated differently, an attachment to freedom of healthcare choice and medical pluralism in a context of a mixed wellness and medical culture. Accessing NPIs is above all unique, the result of a meeting between a life event and a meaningful solution. Some principles need to be remembered:

• Do not rely solely on your intuitions and preferences to choose an NPI, and especially beware of the firstcomer!

• Choosing an NPI is never by chance. It is important to understand the reasons, the circumstances, and the people who advise you. In this personal journey, when symptoms are present, never forget to investigate the possibility of a serious organic or functional hypothesis. Talk to your doctor if you have any doubts or if symptoms persist.

References

Adhanom Ghebreyesus, T. (2019). Foreword. In WHO (Ed.), *WHO global report on traditional and complementary medicine*. Geneva: WHO.

AIR+R. Post-rehabilitation network for patients with respiratory disease www.airplusr.fr.

Apelian, N., & Davis, C. (2018). *The lost book of herbal remedies*. Austin: Capital Printing.

Awortwe, C., Makiwane, M., Reuter, H., Muller, C., Louw, J., & Rosenkranz, B. (2018). Critical evaluation of causality assessment of herb-drug interactions in patients. *British Journal of Clinical Pharmacology, 84*(4), 679–693. https://doi.org/10.1111/bcp.13490.

Centre Fédéral d'Expertise et des Soins de Santé de Belgique. (2011). *Dementia: Which non-pharmacological interventions?* (p. 160C). Brussels: KCE reports.

Chou, W. S., Hunt, Y. M., Beckjord, E. B., Moser, R. P., & Hesse, B. W. (2009). Social media use in the United States: Implications for health communication. *Journal of Medicine Internet Research, 11*(4), e48. https://doi.org/10.2196/jmir.1249.

D'Ormesson, J. (2019). *Je dirai malgré tout que cette vie fut belle*. Paris: Gallimard.

Di Sarsina, P. R. (2017). The social demand for a medicine focused on the person: The contribution of CAM to healthcare and healthgenesis. *Evidence-based Complementary and Alternative Medicine, 4*(S1), 45–51. https://doi.org/10.1093/ecam/nem094.

Dilhuydy, J. M. (2003). L'attrait pour les médecines complémentaires et alternatives en cancérologie: une réalité que les médecins ne peuvent ni ignorer, ni réfuter. *Bulletin du Cancer, 90*(7), 623–628.

DTX Alliance. (2018). *Digital therapeutics: Combining technology and evidence-based medicine to transform personalized patient care*. Brussels: DTX Alliance.

Eisenberg, D. M., Kessler, R. C., van Rompay, M. I., Kaptchuk, T. J., Wilkey, S. A., Appel, S., et al. (2001). Perceptions about complementary therapies relative to conventional therapies among adults who use both: Results from a national survey. *Annals of Internal Medicine, 135*(5), 344–351. https://doi.org/10.7326/0003-4819-135-5-200109040-00011.

European Chronic Disease Alliance. (2019). *Towards an EU strategic framework for the prevention of non-communicable diseases (NCDs)*. Brussels: ECDA.

Fried, L. P., Tangen, C. M., Waltson, J., Newman, A. B., Hirsch, C., Gottdiener, J., Seeman, T., Tracy, R., Kop, W. J., Burke, G., & McBurnie, M. A. (2001). Frailty in older adults: Evidence for a phenotype. *Journal of Gerontology, 56*(3), 146–156. https://doi.org/10.1093/gerona/56.3.m146.

Gray, J. (1992). *Men are from Mars, women are from Venus*. New York: HarperCollins.

Guétin, S., Portet, F., Picot, M. C., et al. (2009). Effect of music therapy on anxiety and depression in patients with Alzheimer's type dementia: Randomised, controlled study. *Dementia and Geriatric Cognitive Disorders, 28*, 36–46. https://doi.org/10.1159/000229024.

Guétin, S., Giniès, P., Siou Kong, A., et al. (2012). The effects of music intervention in the management of chronic pain: A single blind, randomized, controlled trial. *Clinical Journal of Pain, 28*, 329–337. https://doi.org/10.1097/AJP.0b013e31822be973.

Harari, Y. N. (2017). *Homo Deus. A brief history of tomorrow*. Canada: McClelland & Stewart.

Health Canada. (2001). *Perspectives sur les approches complémentaires et parallèles en santé, ministère fédéral de la Santé*. Montreal: Sante Canada.

Healthcare Fraud Prevention Partnership. (2018). *Examining clinical laboratory services: A review by the healthcare fraud prevention partnership*. Baltimore: HFPP.

Higham, S., & Bernstein, L. (2017, October 15). *The drug industry's triumph over the DEA*. Washington Post. https://www.washingtonpost.com/graphics/2017/investigations/dea-drug-industry-congress/

Kuchenbaecker, K. B., Hopper, J. L., Barnes, D. R., et al. (2017). Risks of breast, ovarian, and contralateral breast cancer for BRCA1 and BRCA2 mutation carriers. Journal of American Medical Association, 317(23), 2402–2416. doi: https://doi.org/10.1001/jama.2017.7112.

Li, Y., Pan, A., Wang, D. D., Liu, X., Dhana, K., Franco, O. H., Kaptoge, S., Di Angelantonio, E., Stampfer, M., Willett, W. C., & Hu, F. B. (2018). Impact of healthy lifestyle factors on life expectancies in the US population. *Circulation, 138*(4), 345–355. https://doi.org/10.1161/circulationaha.117.032047.

Lognos, B., Carbonnel, F., Boulze-Launay, I., Bringay, S., Guerdoux-Ninot, E., Mollevi, C., Senesse, P., & Ninot, G. (2019). Complementary and alternative medicine in patients with breast Cancer: An exploratory study of social network forums data. *Journal of Medical Internet Research Cancer, 5*(2), e12536. https://doi.org/10.2196/12536.

Mas, S., Quantin, X., & Ninot, G. (2015). Barriers to, and facilitators of physical activity in patients receiving chemotherapy for lung cancer: An exploratory study. *Journal of Palliative Care, 31*(2), 89–96. https://doi.org/10.1177/082585971503100204.

Melvin, C. L., Dolan-Mullen, P., Windsor, R. A., Whiteside, H. P., Jr., & Goldenberg, R. L. (2000). Recommended cessation counselling for pregnant women who smoke: A review of the evidence. *Tobacco Control, 9*(3), 80–84. https://doi.org/10.1136/tc.9.suppl_3.iii80.

Molassiotis, A., Fernández-Ortega, P., Pud, D., et al. (2005). Use of complementary and alternative medicine in cancer patients: A European survey. *Annals of Oncology, 16*(4), 655–663. https://doi.org/10.1093/annonc/mdi110.

Moullec, G., Ninot, G., Varray, A., Hayot, M., Desplan, J., & Préfaut, C. (2008). An innovative maintenance follow-up program after a first inpatient pulmonary rehabilitation. *Respiratory Medicine, 102*(4), 556–566. https://doi.org/10.1016/j.rmed.2007.11.012.

Music Care. (2020). www.music-care.com

Myint, P. K., Smith, R. D., Luben, R. N., Surtees, P. G., Wainwright, N. W., Wareham, N. J., & Khaw, K. T. (2011). Lifestyle behaviours and quality-adjusted life years in middle and older age. *Age and Ageing, 40*, 589–595. https://doi.org/10.1093/ageing/afr058.

Peeling, P., Castell, L. M., Derave, W., de Hon, O., & Burke, L. M. (2019). Sports foods and dietary supplements for optimal function and performance enhancement in track and field athletes. *International Journal of Sport Nutrition and Exercise Metabolism, 29*(2), 198–209. https://doi.org/10.1123/ijsnem.2018-0271.

Prince, M., Bryce, R., & Ferri, C. (2011). The benefits of early diagnosis and intervention: World Alzheimer report. In *Alzheimer's disease*. Washington, DC: Alzheimer's Disease International.

Pujol, P., Vande Perre, P., Faivre, L., et al. (2018). Guidelines for reporting secondary findings of genome sequencing in cancer genes: The SFMPP recommendations. *European Journal of Human Genetics, 26*(12), 1732–1742. https://doi.org/10.1038/s41431-018-0224-1.

Rakel, D., & Weil, A. (2018). Philosophy of integrative medicine. In *D. Rakel integrative medicine* (4th ed.). Philadelphia: Elsevier.

Scrivener, L. (2000). *Terry fox: His story*. Toronto: McClelland and Stewart.

Servan-Schreiber, D. (2011). *Anticancer: A new way of life*. London: Pinguin Group.

Steptoe, A., Deaton, A., & Stone, A. A. (2015). Subjective wellbeing, health, and ageing. *Lancet, 385*(9968), 640–648. https://doi.org/10.1016/S0140-6736(13)61489-0.

Stuckelberger, A. (2008). Anti-ageing medicine: Myths and chances. Zurich: vdf Hochschulverlag AG.

Thompson, M. J., Laver-Bradbury, C., Ayres, M., et al. (2009). A small-scale randomized controlled trial of the revised new forest parenting programme for preschoolers with attention deficit hyperactivity disorder. *European Child Adolescent Psychiatry, 18*(10), 605–616. https://doi.org/10.1007/s00787-009-0020-0.

Tourette-Turgis, C. (2020). https://universitedespatients-sorbonne.fr/

WHO. (2013). *WHO traditional medicine strategy 2014–2023*. Geneva: WHO.

WHO. (2017). *Global strategy and action plan on ageing and health*. Geneva: WHO.

Yao, S., Liu, Y., Zheng, X., et al. (2020). Do nonpharmacological interventions prevent cognitive decline? A systematic review and meta-analysis. *Translational Psychiatry, 10*(1), 19. https://doi.org/10.1038/s41398-020-0690-4.

Young, S., & Amarasinghe, J. M. (2010). Practitioner review: Non-pharmacological treatments for ADHD: A lifespan approach. *Journal of Child Psychology and Psychiatry, 51*(2), 116–133. https://doi.org/10.1111/j.1469-7610.2009.02191.x.

Chapter 7
The Market for Non-pharmacological Interventions

1 Introduction

The world market for non-pharmacological interventions (NPIs) has been increasing for 10 years, led by countries such as China, Germany, India, Japan, Switzerland, and the United States. The use of NPIs concerns all generations, ages, sexes, and territories, across the range of socioeconomic status and all sociocultural levels. The demand is large, from wellness to therapy, from health prevention to biomedical product potentiation, from wellness enhancement to healthy aging, from touristic centers to hospitals. The increase in NPI supply is in the process of structuring itself. This chapter details the characteristics of a major and disruptive market organized in a new ecosystem and a win-win alliance between classical and recent actors in the health-related market.

2 Consumers

Ten years ago, the typical consumer of NPIs was a 34-year-old working woman, middle class or higher, living in a household of one to three persons. She wanted to get away from ubiquitous chemicals and take care of herself naturally, both in terms of beauty and health as well as fitness. In Switzerland, for example, the user profile in 2012 was more frequently female, of middle age, and with higher education (Klein et al. 2015). Today, this interest has expanded to include all sexes, ages, and territories, across the range of socioeconomic status, and all social categories. About 66% of people use NPIs.

Some disease treatments such as oncology require NPIs (1) to reduce the side effects of biomedical treatment; (2) to increase health status; and (3) to improve quality of life. About 80% of patients use NPIs, but many do not mention this to their oncologist, contributing to the difficulty of obtaining accurate frequency of NPI use (Molassiotis et al. 2005).

The consumption of NPIs is encouraged by evidence of their effectiveness to improve certain functions of the body, to compensate for certain deficiencies (for example, in vitamin D), or to reduce the risks of chronic disease. However, many NPIs are adopted long before science has established their safety and efficacy (Jonas and Levin 1999). This differs from biomedical treatment procedures usually introduced by professional corporations and industries rather than by the public. The public adopts and seeks out these practices first, and healthcare professions and industries follow. *"This says something about the changing nature of public preferences and professional responsiveness to those preferences"* (Jonas and Levin 1999, p. 3). Thus, "unconventional" practices will arise from specific groups, will become more "professionalized" and adopted into the mainstream.

2.1   The Reasons

There are many reasons why consumers have been encouraged in the use of NPIs in the last decade, sometimes simultaneously, sometimes consecutively, often relevantly, always passionately:

- Preferring natural products and human services perceived as safer for health and well-being
- Living in peace without systematic testing and anxiogenic medical procedures
- Improving physical attractiveness
- Enhancing performance
- Strengthening defenses
- Healthy aging
- Preventing recurrent health conditions (e.g., fatigue, flu, cold, allergy, seasonal depression) and disease (acute or chronic disease)
- Decreasing biomedical side effects and resistances (e.g., adverse effect, toxicity, sequelae, treatment failure, lack of responsiveness, persistent pain)
- Maximizing chances of recovery
- Improving quality of life

From the "Patient" Patient to the Empowered and Engaged Consumer
The new century has created humans aspiring to be actors in their care pathways and no longer to be passive as suggested by the word "patient" (Chap. 6). They want to reclaim their bodies and try to find meaning in their illness. *"Pressures on health care's traditional break-fix model to change—to become more proactive, predictive, and focused on well-being—are coming from multiple directions, including a shift in attitudes and behaviors toward greater consumer engagement and empowerment. No longer passive participants in the health care process, consumers are demanding transparency, convenience, access, and personalized products and services—which they get in all other aspects of their life."* (Allen 2019, p. 13).

Whatever the individual motivations, a large majority of individuals turn to NPIs in addition to authorized biological treatments or preventive interventions. The choices are made with discernment, contrary to what is suggested by some in the uninspired and unsympathetic mass media. When the disease is serious, patients do not avoid conventional medicine. In the case of cancer, for example, it is rare for patients to consult an unconventional therapist before a physician (Dilhuydy 2003). Only 4% of Europeans exclusively use alternative medicines, while more than 66% use them as supplements to better support medical treatments and to increase their chances of recovery (Molassiotis et al. 2005). This evolution has led to the creation of a specific term: "supportive care" or "integrative care." The scientific *Journal of the National Cancer Institute* devoted a complete monograph to integrative care in 2017 (Mao 2017). A learned society was even created in 2003, the *Society for Integrative Oncology*, with the ambition to promote evidence-based supportive care to improve the lives of people living with cancer.

Most users of NPIs feel satisfied, although some recognize that the methods used do not always work and that they sometimes have had to change therapists before finding a satisfactory one for them (Begot 2008). They consider that the feeling of relaxation and well-being is as important as the therapeutic result (Centre de Recherche et d'Information des Organisations de Consommateurs 2012).

The increase in demand for NPIs reflects a disruptive ambition on the part of all generations to be treated differently with an emphasis on freedom of choice and medical pluralism (Graz et al. 2011) in the context of hybridization between physical health and well-being. The mission is to defend life in a very pragmatic sense at a time when health is the priority for the population. The worldwide experience of COVID-19 attests to this evolution. Such an approach focuses on the human being in his or her wholeness, integrity, and full dignity to enable a suitable and free choice of the individual health program (Di Sarsina 2017).

2.2 Rates

A phone survey in 2007 with a representative sample of the French population of 958 people aged 18 years and older found that 39% of people use "natural medicines" (IFOP 2007). The vast majority considered these solutions effective for prevention (78%), to treat stress related to illnesses (77%), or to treat mild illness severity (76%). With respect to dietary supplements, for example, 41% of cases were following a medical prescription. In 15% of cases, it was the advice of a loved one or a pharmacist that triggered consumption. The discovery of a product on a shelf in a shopping center or on the Internet also led to the purchase in some cases. These supplements were mainly consumed by artisans-traders-entrepreneurs (23%), by workers (17.5%), and by professionals-managers (17.5%). In 2012, the Center for Strategic Analysis estimated that 66% of French people used "complementary and alternative medicine." This rate was lower than that for Germany and northern European countries.

According to the European research program CAMBrella (2012), more than 100 million Europeans use complementary and alternative medicine. This rate remains lower, however, than that for North American and Japanese citizens.

3   Practitioners

According to the CAMbrella survey carried out in 2012, 150,000 European physicians exclusively practiced complementary and alternative medicine, like acupuncturists. These practitioners are curious and open-minded. They declared themselves to be different from allopathic and reductionist practitioners (e.g., organ experts like cardiologists or method experts like knee surgeons). NPI practitioners affirmed working with attentiveness and availability (Simon et al. 2007). They made extensive use of knowledge from psychosomatic medicine, which has created a stronger awareness of the power of the mind to influence bodily function (Chap. 3). The notion that mind and body are fully integrated and that emotional states affect health has been foregrounded in integrative medicine (Jonas and Levin 1999). Combined mind-body interventions incorporate various forms of energy and mind exercises to heal ailments. Yoga has been dominant in the world market for NPIs before 2010, whereas hypnotherapy and many guided practices have grown in the last decade (e.g., MBSR).

Since 2010, more and more classical physicians have become interested in NPIs, both in terms of prescription and their own practice. This is particularly the case for general practitioners and chronic disease specialists wishing to propose a global approach, personalized optimization, and a multimodal strategy to their patients. They want to respond to complex patient needs (e.g., a patient with several comorbidities) by combining therapies and preventive strategies. This is the case for internal medicine physicians, physical medicine and rehabilitation physicians (also known as physiatrists or physiotherapists), pain physicians, psychiatrists, nutritionists, palliative medicine physicians, and sports medicine physicians. NPI practice has been increasing among young general practitioners, perhaps because of the feminization of the field and because of their own personal practice preferences (e.g., methods for stress reduction, concentration improvement, decrease in fatigue). They have experienced benefits for themselves as they cope with ever-increasing workloads and responsibilities.

The 2012 CAMbrella survey identified 178,000 paramedical professionals in Europe. They belonged to traditional health disciplines. At the same time, these professionals had acquired qualifications related to each NPI. Classical physiotherapists become osteopaths or chiropractors; pharmacists become herbalists or mycologists; dieticians become lifestyle coaches; nurses become therapeutic educators or online case managers; and occupational therapists become consultants on healthy aging.

Prevention, education, and social professionals are also joining the movement. Psychotherapists, teachers in adapted physical activity, and social workers offer their services.

And as if that were not enough, newly entitled professions are emerging with the suffix "-path" (e.g., naturopath) or "-ist" (e.g., musicologist). To date, it is impossible to know precisely the number of trades and professionals, to know their real activity, their diploma or qualification, and their initial and continuing training (Chap. 9).

But that is not all, far from it. Traditional medicine practitioners now display professional plaques outside their offices in the street and no longer on pieces of paper slipped into your mailbox: European healers, African marabouts, Chinese therapists, Ayurvedic practitioners.

Practitioners have been stimulated by the continuous growth of the NPI market during the last decade, particularly in Europe, which is operated within a weak regulatory regime (WHO 2001, 2019). Governments do not require practitioners to be licensed in order to operate a business, so almost anyone with the inclination can establish a practice. The majority of practitioners work on a freelance basis. They can operate a weekly on-site treatment program for employees in companies and for private persons. Some choose to work in a dedicated integrative medicine center or in a conventional medical center.

4   Centers

NPI practitioners, whether medical doctors or not, for a long time were isolated in city offices or specialized activities (e.g., palliative care, addiction, mental disorder, geriatrics, Thermalism), have recently been integrated into hospital teams and departments. French institutions are particularly advanced, such as the Pitié-Salpêtrière hospital in Paris for the treatment of pain.

The Growth of Thermalism in France

- In 2015, spa treatments accounted for 390 million euros in healthcare consumption (medical consultation, paramedical care, daily packages).
- France represents 20% of European thermal provision, with 89 stations and 105 thermal establishments (Atout France 2015).
- Hydrotherapy grew 6.4% between 2013 and 2014.
- 77.8% of spa guests come to treat or relieve rheumatological problems, 7.8% for respiratory problems.
- Treatments reimbursed by health insurance represent nearly 90% of the turnover of thermal establishments.

Hospitals are making progress in building more efficient services based on an integrative and human-oriented approach. Early detection and the anticipation of relapses make it possible to limit heavy treatments and therefore expenses. The personalization of prevention and care activity on a large scale makes it possible to allocate resources in a targeted way to limit the additional costs. The particular healthcare strategy depends on the patient's ability to cope with the disease diagnosis, the cascade of analyses and treatments to which he or she will be subjected, and the patient's social support system, with the aim of a return to an active social and professional life after the illness. In cancer treatment, for example, some North American centers have taken the turn toward inclusion of integrative therapy, with the creation of integrative therapy departments such as the Dana-Farber Cancer Institute (Boston, Massachusetts), the MD Anderson Cancer Center (Houston, Texas), the Memorial Sloan Kettering Cancer Center (New York, New York), and the UCSF Osher Center for Integrative Medicine (San Francisco, California). Some European centers are proactive such as the Karolinska Institute (Stockholm, Sweden), the University College Hospital Macmillan Cancer Centre (London, UK), and the Christie NHS Foundation Trust (Manchester, UK). To mention a few examples in France, the Center Léon Bérard (Lyon), the Institute Gustave Roussy (Paris), the Institute Raphael (Paris), and the Montpellier Cancer Institute (Montpellier) are very active in the NPI field. The Montpellier Cancer Institute has a number of innovations, for example, in proposing patient programs of art therapy, exercise, serious games, personalized dietetics, psychotherapy, disease management education, and hypnotherapy. NPIs are becoming integral in the healthcare system in Asia, especially in China, where most of the hospitals in the country have distinct departments and wards that are dedicated to traditional therapeutic approaches.

And this is just the beginning. Multi-professional health centers with private-public contracts are deployed in each town to meet the multifocal demands of patients after hospital discharge or for health prevention, at a proximal distance from their homes, (e.g., integrative health institute, health house, home family holistic center).

5   Producers

The demand for herbal remedies is expanding globally, beyond the known uses in traditional medicines (e.g., traditional Chinese, Indian, and African Medicines). Plant-derived products with minimal or no industrial processing have been used to treat disease in local or regional healing practices. Herbal products, also called botanicals, herbal medicines, or phytomedicines, are available in all forms (e.g., dried, chopped, powdered, or liquid), come from different origins (e.g., herb, plant, tree bark, mushroom, seaweed), and can be used in various ways (e.g., swallowed as pills or powders; brewed as extract beverage or tisane; applied to the skin as gels, oils, lotions, or creams; added to bath water). Economists pay significant attention to botanicals. This NPI segment has dominated the global market. Eighty percent of

African populations use some form of traditional herbal medicine (Willcox and Bodeker 2004), and the worldwide annual market for these products approached US$60 billion in 2001 (WHO 2002). Herbal supplement sales in the United States experienced record growth in 2018, increasing by an estimated 9.4% since 2017 (Smith et al. 2019). Consumers spent a total of $8.842 billion on herbal supplements across all market channels in 2018 (Smith et al. 2019). The market in France is estimated at 240 million euros with an increase of 10% per year since 2010. Producers include Nordic Nutraceuticals, Pure encapsulations, AYUSH Ayurvedic Pte Ltd, and Sheng Chang Pharmaceutical Company. They have created strategic partnerships with drug industries to establish their presence in the market. For instance, in June 2019, Luye Pharma Group partnered with AstraZeneca to formulate a red yeast rice product to assist patients suffering from cholesterol. Novartis, Sanofi, and Pierre Fabre have also developed partnerships, especially to expand their activities.

Sales of Mushrooms
Mushrooms were primarily sold in the form of vegetable capsules and powders. In 2018, the *"sales of mushrooms increased by 40.9% from 2017 and totaled US $7,800,366"* (Smith et al. 2019, p. 71).

The food supplements market is also growing by 5.3% per year (Fig. 7.1). In 2016, it reached 1.62 billion euros in France. These products are distributed through various channels including pharmacies (52% of sales), specialized outlets, large and medium-sized stores, and the Internet. Pharmacies sell food supplements primarily for:

Fig. 7.1 Many people take food supplements. (Published with kind permission of © Plateforme CEPS. All rights reserved)

- Tone (also called "vitality"), worth 142 million euros (annual growth rate of 8.1%)
- Transit (also called "digestion"), worth 130 million euros (annual growth rate of 20.2%)
- Improving sleep and reducing stress (also called "relaxants"), worth 124 million euros (annual growth rate of 15.1%)

Cosmeceuticals follow the same dynamic (Martin and Glaser 2011). Sun protection creams and skincare creams represented a market of 6.6 billion euros in France in 2009. The desire to look good, younger, toned, and healthy has caught on across the global population in a big way. According to Euromonitor International, total global retail sales of natural-product-based cosmeceuticals was US$2.98 billion in 2015, with this segment showing a 4% compound annual growth rate during the previous 5 years. Given the growing interest in these products among patients and the strong claims made by manufacturers, the Indian cosmeceuticals market has been showing a consistent growth between 15% to 20% per annum since 2010 (Pandey et al. 2020).

Manufacturers and sellers of sports equipment are increasingly turning to emphasizing health in the design and marketing of their products. They adapt their equipment to all ages and levels. The innovations are considerable. Decathlon, a major French brand that is both a manufacturer and a distributor, even goes so far as to develop and share a free mobile application with a voice coach, a performance telemetry system, and a program for planning training sessions ("*Decathlon coach*"). Obviously, recommendations are made in terms of equipment, clothing, food, drink, and all kinds of accessories.

6   Decision-Makers

Far from a countercultural loophole or a beatnik's revival, decision makers aim to respond to aging populations and rising patient expectations in the context of substantial budget constraints for health care and social support. The sustainability of the national health system mobilizes all governments. Patient-centric, data-driven, and preventive approaches have become decisive elements in the delivery of sustainable health and social care systems, and to improve the lives of populations.

The Time Has Come to Shift the Healthcare System
"It is crucial for health care stakeholders to work toward a future in which the collective focus shifts away from a system of sick care—treating patients after they fall ill—to one of health care, which supports physical and mental well-being, prevention, and early intervention. Some steps are being taken in this direction. In partnership with sectors such as employment, housing, education, transportation, retail, banking, and technology, health care stakeholders are beginning to employ a smart health community approach to collectively help drive innovation, increase access and affordability, improve quality, and lower costs through more efficient delivery models." (Allen 2019, p. 4)

Some private health insurance plans offer full or partial coverage for NPI treatments such as osteopathy, chiropractic, and acupuncture, while 90% of alternative medicines are purchased privately (Thomas et al. 2001).

7 New Market Players

7.1 Tourism Industry

NPIs are a booming business for the tourism industry. Hoteliers and accommodation centers offer NPIs in their recently installed spas and wellness areas at the request of their customers.

Europe is attractive in terms of the number of wellness travelers; in fact, it is the world's leading destination. In 2012, 216 million travelers came to Europe, compared to 172 million to North America and 152 million to Asia. Europe has 32,000 spas, some of which are attached to an accommodation facility. France ranks fifth in the world in terms of attendance at health tourism venues and fourth in terms of revenue, ahead of Austria, and behind the United States, Germany, and Japan (Global Wellness Institute 2017). Health tourism is expanding in China, Thailand, Singapore, and other Southeast Asian countries. For instance, the Hainan Chinese province focuses on providing high-end health services, including cancer and cardiovascular disease treatment and rehabilitation, health, and wellness programs using both cutting-edge technologies and traditional Chinese medicine, cosmetic services, and retreat tourism.

Thalassotherapy (the use of seawater in cosmetic and health therapy) is not to be outdone (Fig. 7.2). An AFNOR standard was recently created in France for this therapy (XP-X50-844), for the 55 centers run mainly by the Accor, Thalazur, and Lucien Barrière groups. Their turnover fluctuates between 125 and 130 million euros for around 400,000 "spa guests" per year (Atout France 2015). Balneotherapy (e.g., immersion in mineral water or mineral-laden mud), more often linked to the leisure sector, is part of this overall movement.

> **A Development Focused on Wellness and Health Tourism in France**
> Created in 1996, a development of wellness tourist facilities included 485 spas in France in 2015 with 3.5% of the hotel fleet (48% in 5-star hotels, 17% in 4-star hotels). The development is notable for its tourist residences, village clubs, holiday villages, and campsites. *"We need to focus more on health tourism. A popular destination for tourists, France has developed spa towns, spas and hotel infrastructures. It must promote its regions as destinations for well-being and relaxation through the prism of health, especially since this sector is not likely to destabilize the public healthcare system. There is a sector here that will become attractive, with aging, chronic diseases, health problems linked to stress and the growing desire to prevent diseases"* (Marguerit and Reynaudi 2015, p.8).

Fig. 7.2 The NPI offer in
thalassotherapy centers,
spas, hotels, and campsites.
(Published with kind
permission of ©
Plateforme CEPS. All
rights reserved)

7.2   Distribution Industry

Large-scale distribution has characterized the NPI market after the success of pioneers and their innovative distribution models (e.g., Herbalife). Supermarkets are starting to sell parapharmaceuticals, food supplements, herbal medicines, minerals, mushroom preparations, essential oils, cosmeceuticals, and digital health interventions at the head of the aisles with their proactive discount method. Large supermarkets install complementary medicine and wellness zones in their stores. Small brands are following the movement, for example, by highlighting sales areas for books and magazines on personal development, wellness, and health.

> **Worldwide Success Story of Herbalife**
> Herbalife with their independent distributors operates in more than 90 countries around the world with net sales of US$4.5 billion in 2015. The company aims to help people achieve good nutrition and a healthy and active life.

The story is only in its infancy. The products sold are accelerators of information dissemination on NPIs. Discounters and teleshopping services make their know-how available to distributors and manufacturers to expand the market. The demand is so important that the e-commerce giant Amazon is now entering the healthcare sector. Amazon has been searching for solutions targeting specific uses (e.g., chronic disease management) and population demands (e.g., health products delivery). The company has been reinvesting in massive infrastructure that will be a significant advantage as it enters health care. Amazon focuses on user and consumer experiences to persuade them to switch to its offerings. Without the need to make money in health care, Amazon can take advantage of the high margins and convoluted components of the healthcare business that are ready for disruption. Despite various obstacles, including established processes, market leaders, and a general risk-aversion of buyers to new players, the entrance of Amazon in the healthcare sector will change how the system is designed or force existing players to become more competitive.

8 Startups

Startup creators have grasped health issues and the associated looming revolutions on the horizon. The potential for innovation is enormous, particularly in the wellness and personal services sector, which is far less regulated than that for drugs and medical devices.

These innovative and disruptive companies are creating new non-pharmacological solutions. They combine them and associate them with existing drugs or medical devices. They embellish them with great marketing strategies. They protect their innovation, their brand, their algorithm, and their know-how.

Supported by incubators and health clusters ready to do anything to get them off the ground, and investors who are not very attentive, they free themselves from old-fashioned shackles and raise considerable funds. They risk everything. They see far and wide. They break codes, and not just dress codes. They no longer count their working hours, especially since they are dedicated to developing concrete actions for the benefit of human health.

They recruit goodwill, from the trainee to the family member, through the senior manager in retraining. They weave intergenerational ties. They bring the professions together. They hybridize worlds. They adjust to demand with disconcerting agility. They help each other and compete with each other. They are inspired by foreign successes.

Success Stories of Some Startups
Some French startups stand out, with personalized solutions for patients with chronic disease or elderly people using videoconferencing (e.g., Mooven), actimetry technologies (e.g., Siel Bleu), or serious games (e.g., Natural Pad). They are found in horse therapy (e.g., Equiphoria). Manufacturers of food supplements such as spirulina (e.g., PhycoBiotech) or essential oils (e.g., Sirius) are now exporting worldwide. Some develop innovative health solutions in music therapy (e.g., Musicare) or electronic pill boxes (e.g., La Valeriane). Some help with artificial intelligence technology to choose the best NPIs (e.g., Kalya).

9 Multinational Companies

9.1 Drug and Medical Device Industries

At the beginning of the *iCEPS Conference* in 2011 (an international scientific congress on the scientific, medical, legal, business, and ethical questions raised by NPIs), we thought that the drug and medical device industries would be hostile to the NPIs field, considering them as alternatives to their products. Actually, not at all.

They came to listen to presentations by international experts, including the results of rigorous studies by researchers around the world and testimonies from users. They understood that NPIs complement their products and, in some cases, potentiate them. Some large drug companies are developing NPIs today, conceived as an associate service to their products.

NPIs: An Essential Complement
Biomedical industries share the same vision as NPI manufacturers and practitioners: improve health and quality of life of people using a scientific method and a quality approach. They have understood that NPIs do not compete with their products, but on the contrary complement them to improve and lengthen the lives of citizens and assist people with health risks.

Since then, they have participated in the *iCEPS Conference*, supporting research in the field. Some even develop NPIs. Food manufacturers market food supplements and health foods. Pharmaceutical manufacturers offer complementary services to individuals, such as connected pill dispensers. Digital manufacturers are developing disease management solutions.

Of course, not all of them are convinced yet, especially when considering the views of their shareholders. We can suspect that some of them want to enhance their own good conscience or to improve their image after the scandals of the past century. But never mind, the whole movement is manifest and deep. In the end, the informed consumer will end up being right.

9.2 New Technology Industries

Many technology industries are advancing in the health area by playing to their strengths: Apple's patient-centric vision prioritizes consumers, Google applies artificial intelligence to everything from drugs to lifestyle management solutions, Microsoft builds a cloud platform of health data management, IBM works on the Watson health decision machine, Facebook adds a healthcare feature to its platform to provide preventative healthcare information such as suggested checkups, and Twitter encourages relationships between healthcare companies and patients or customers. These global and multi-billion-dollar companies hope to find better and more sustainable ways for creating healthier communities. They also help traditional medical companies to make the transition from being a provider of products to a vector of experiences, services, and solutions. Health clinics, on-site experts offering diet, nutrition, and other advisory services are becoming popular differentiators.

10 New Private Investors

10.1 Free Beta-Testers

NPIs have become an experimental sandbox for investors. They have excellent beta-testers at low cost, ready to try anything and everything: athletes, body-builders, extreme explorers, astronauts, soldiers, professional gamers, anti-aging champions, patients at the end of their lives, and bloggers. They only have to dig into the relevant uses of "real life" with big data analyses to identify the nuggets of tomorrow. Some may become new drugs or new medical devices, if it is worth the effort.

10.2 Opportunities from Chronic Diseases

States have taken stock of the social and economic problems caused by chronic noncommunicable diseases and have called on companies to help them cope with the massive burden (Chap. 2). Some countries have fallen behind on this ambitious roadmap, particularly in France. According to the OECD (2016), prevention is a "*weak link*" in France: only 2% of current health expenditure is devoted to it, compared to 3%, on average, in Europe. This proportion is insufficient. Preventive practices will undoubtedly increase on the part of the public authorities as well as the private sector, favoring the establishment of NPIs with the aim of reducing harmful health behaviors: first, sedentary lifestyle, excessive consumption of alcohol, smoking, and poor diet. A national health prevention strategy was launched in 2018. Primary prevention concerns healthy people, but prevention is also very useful for persons with disease. In this case specialists use the terms secondary and tertiary prevention. Investments will be made in this area, with regard to increased screening, reducing unnecessary health expenditure (transporting patients, for example), or avoidable problems (misdiagnosis, surgical errors, or drugs taken incorrectly by patients). In 2014, drugs dispensed in city pharmacies to people over the age of 60 cost € 13.4 billion, 83% of which was reimbursed by the various health insurance plans, i.e., € 585 and totaling 63 boxes on average, per person, per year (Morin and Laroche 2016). Healthcare use that did not comply with medical prescriptions cost health insurance companies 2 billion euros, including 1 million days of hospitalization. NPIs chosen for their benefits and low cost are bound to be associated with biomedical treatments.

10.3   Finance

With new technologies making the traceability and predictability of human behavior a reality, insurance companies (e.g., AG2R Allianz, AXA, Generali), banks, and wealthy foundations (e.g., disease alliances, Bill & Melinda Gates Foundation, Qatar health foundation) have been investing in NPI infrastructure, training, delivery, and post-marketing surveillance. This investment in the future of human well-being is more noble than investment in the military or fossil energy industries, is more attractive than in chemical industry activities, is more ethical than in polluting industries, and is more concrete than trading on the stock exchange. Investing in health and human well-being is a way of avoiding conflicts between communities and wars between states, as we know what they cost in the last century. The media make no mistake about it given the daily number of articles and advertisements.

> **NPIs: An Advertising Domain with Unlimited Possibilities**
> Advertising regulations do not exist in the NPI field, unlike the very restrictive rules for drugs and medical devices. The same is true for NPI trading, which is not subject to the public health code governing health professionals. Because NPIs interest public opinion and because their actors can finance advertising inserts, the terms "alternative medicines," "natural remedies," "complementary therapies," "integrative health," "positive health," and "traditional medicines" are more and more visible. The famous celebrity magazine *Paris Match* published a special issue in January 2017 on natural medicines. Dedicated health and wellness magazines are increasing. Television reports and debates are widely followed. The French program by Michel Cymes and Adriana Karembeu, devoted to complementary medicines, that was broadcast on March 7, 2017 on France 2, had 3.5 million viewers, or 14.9% of audience share. The documentary series *"Goop Lab"* produced by Gwyneth Paltrow, giving information on well-being and health, is a success on Netflix. The messages on "cutting-edge wellness advice" and "vetted travel recommendations" get through. The curated shop is very close.

11   Public Investments

As detailed previously (Chap. 2), chronic diseases collectively account for over 86% of deaths in the European Union. Almost half of chronic disease deaths occur prematurely, in people under 70 years of age. One-quarter of all chronic disease deaths occur in people under 60 years of age. Beyond the human cost to society, this alarming data directly affects national productivity and income.

This situation affects states' activities with high direct and indirect costs related to early death and absenteeism.

> **The Burden of Noncommunicable Diseases**
> *"Work-related annual costs of noncommunicable diseases to the European economy add up to €610 billion per year, including costs to employers, lost economic output and costs to social welfare systems. Overall, 1.7% of Gross Domestic Product in the EU is spent on disability and paid sick leave each year, which is more than the expenditure on unemployment benefits. This while a healthy workforce is essential to support an ageing population"* (European Chronic Disease Alliance 2019).

Moreover, poverty is a powerful contributor to chronic diseases. Preventing these diseases reduces health inequalities by narrowing gaps between the vulnerable and privileged populations. Health represents a strong economic sector, is a source of employment for workers, is a driver of innovation in research and production, and directly results in growth for the economy. NPIs, because they require time on the part of the healthcare provider to interact with patients, become a sector that provides employment and creates new professions. Public decision-makers are not mistaken. The French National Authority for Health claims the importance of *"improving access to the supply of non-pharmacological therapies"* (French National Authority for Health 2011, p. 52). New jobs are being created in:

– Upstream assessment (e.g., the city of Strasbourg has successfully experimented with prescription sports where specialists in physical health activities offer an assessment of resources and an individual motivational strategy for chronically ill patients in agreement with their treating physicians)
– Counseling (e.g., in pharmacies for NPIs, consulting, and disease management programs)
– Screening (e.g., early detection of neurodegenerative diseases)
– Follow-up (e.g., online case manager for fragile people)
– Surveillance (e.g., regional health network, analyst, and strategic advisor on NPIs)

11.1 Research and Innovation

Since 1998, the United States has invested $2.5 billion in an institute dedicated to research and development of NPIs, the National Center for Complementary and Integrative Health (NCCIH).

Europe contributes sporadically through calls for tenders for research on health innovation and aging well, such as the Horizon 2020 program. The CAMBrella research project coordinated by Wolfgang Weidenhammer in Germany between January 2010 and December 2012 was part of this European program (FP7-HEALTH-2009-3.1-3, No. 241951).

Germany, England, Australia, Saudi Arabia (Khalil et al. 2018), Belgium, Brazil, Canada, China, Iran, Israel, Japan, Norway, Switzerland, and Turkey are countries that invest in research and raise funds from donors.

France has not been outdone since 2010. It encourages NPI research, in particular with the National Cancer Institute (INCa), the National Institute of Health and Medical Research (INSERM), and the Institute of public health research (IRESP). The Foundation de France, the ARC, the Ligue contre le Cancer, the association France Alzheimer, to name just a few, actively support this research effort. Patient associations are mobilizing to support this research. Companies and their various committees are also developing initiatives. Associations of professionals or patients collect funds for research through various initiatives.

12   Market Size

The size of the global NPI market is difficult to establish with precision at the moment because many economic analyses are still segmented: wellness vs. health, healthier diets and public lifestyle programs, health interventions, personal services vs. medical care, mobile health tools vs. medical devices, nutritional products vs. sports activities, and herbal remedies vs. drugs. The measures are imprecise with declarative surveys under- or overestimating rates, with the absence of an exhaustive list of NPIs (Chap. 1) and with sectoral analyses and untracked practices (e.g., natural healer consultation, traditional medicine's free usage, undeclared practice). Despite a growing interest in various parts of the world and improved tracking, the frequency of use of unconventional therapy is far higher than previously reported, for example, in the United States (Eisenberg et al. 2001). The real size of the NPI market remains vague, also because of heterogeneous health insurance coverage (classic or supplemental), and differences between countries and between socioeconomic statuses.

A Key Role in Economic Development According to the World Health Organization (WHO)
"Traditional and complementary medicine is growing and expanding, particularly with respect to products bought in person or over the Internet. The Traditional and complementary medicine sector now plays a significant role in the economic development of a number of countries. At the same time, with prevailing current global financial constraints, use of Traditional and complementary medicine for health promotion, self-health care and disease prevention may actually reduce health-care costs." (WHO 2013, p. 18).

It is interesting to note that in Switzerland, a large survey (18,357 respondents) representative of the population aged 15 and older living in a private household observed an unchanged usage of complementary medicines between 2007 and 2012. This contrasts with the continuous increase of NPI use. The hypothesis can be made that the market and science jointly manage to support NPIs; in other words, those which are effective and safe, to the detriment of alternative and traditional practices that are ineffective or even dangerous.

Some NPI subsector markets are better known, especially in the case of material products (e.g., herbal preparations, minerals, mushroom preparations, supplements, digital applications, cosmeceuticals), and less so for human services (e.g., exercise programs, manual therapies, psychotherapies, nutritional advice) or derived activities (e.g., training, learning sessions, practice manuals, consulting, case management).

The Structuring of an International Market

At this point in time, having a precise idea of the international NPI market as a whole is complex. Some overlaps exist between, for example, public health interventions, paramedical industry products, drug products, touristic services, educational managements, and social supports (Chap. 1). Some borderline proposals can or cannot be counted such as alternative medicine, cosmeceuticals, medical devices, and novel foods available in some countries. Some are partially or totally reimbursed and thus counted by health insurance. Some are components of conventional medicine instead of belonging to traditional medicines, depending on the country's health culture. Last but not least, some have created partnerships with community stakeholders to extend activities. The ecosystem is going to be clearly differentiated and aggregated from segmented sectors.

The OECD has published excellent reports on current state health spending on drugs, individual medical and paramedical curative care (care, surgery, radiotherapy, rehabilitation, ancillary services, and medical goods). On the other hand, the identification of individual uses of NPIs is complex with patients with chronic disease using multiple treatments. Indeed, *"in the case of long-term care, only the medical aspects are generally considered in health expenditure, although it is difficult in some countries to clearly distinguish the health component and the social component. Certain countries which have set up integrated long-term care systems favoring social services may be misclassified in terms of expenditure according to the Health Accounts System, because these social components are not taken into account"* (OECD 2015, p. 174). To date, the identification of health interventions less well-defined than surgeries seems even highly complex. In any case, the rates are underestimated. The digitalization of health services and usages will improve their accuracy and transparency over the next 10 years to contribute to addressing universal health challenges (Chap. 2).

To get an idea of the size of the market, the more plausible international data comes from two institutions. WHO is estimating a global market of 68 billion euros for alternative and complementary medicines (WHO 2013). The Global Wellness Institute (2017, 2018) for its part estimates US$199 billion in 2015 and US$359 billion in 2017 (Table 7.1), and this without taking into account other associated activities that would push it to US$4.22 trillion in 2017, with average annual growth between 2015 and 2017 of 6.4%. Whatever the size, there is no doubt that the market for NPIs is growing rapidly in the world, and more particularly in Europe, given its sanitary, climatological, tourist, and logistical advantages.

> **A Bubble-Free Market**
> The NPI market is expanding internationally in a sustainable manner because it integrates the forces of current changes: digitalization, globalization, rationalization, personalization, and pluralism of choice in a sector of activity, health, which is not like the others. Moreover, it requires rigorous evaluation to guarantee clients – users – patients the benefits that they expect.

Table 7.1 Statistics of the Global Wellness Institute (2018)

	2013 (billion $)	2015 (billion $)	2017 (billion $)	Year variation 2015–2017
Personal care, beauty, and anti-aging	1015.6	999.0	1082.9	+4.1%
Healthy eating, nutrition, and weight loss	574.2	647.8	702.1	+4.1%
Wellness tourism	494.1	563.2	639.4	+6.5
Fitness and mind-body	446.4	542.0	595.4	+4.8%
Preventive and personalized medicine and public health	432.7	534,3	574.8	+3.7%
Complementary and alternative medicine	186.7	199.0	359.7[a]	[a]
Wellness real estate	100.0	118.6	134.3	+6,4%
Spa	94.0	98.6	118.8	+9.8%
Spa economy	50.0	51.0	56.2	+4.9%
Workplace wellness	40.7	43.3	47.5	+4.8%
Total	3367.8	3724.4	4220.2	+6.4%

[a]The authors of the 2018 report modified the category of complementary and alternative medicines between 2015 and 2017 by adding traditional medicines (traditional Chinese medicine, Ayurveda medicine, for example); the percentage change 2015–2017 is not included because it would not be comparable to the others
Reproduced with permission from the Global Wellness Institute

12.1 Perspectives

All qualitative surveys converge: an increasing demand for and use of NPIs. By taking an interest in the whole individual and in prevention, the medical world is redeploying toward autonomy, healthy aging, and wellness. A positive medicine and a positive health movement complements restorative medicine, where each new innovation is becoming costlier and more restricted to a small part of the population. Many studies showed important cost savings with NPIs. For example, a randomized controlled trial included an economic analysis of physiotherapy, manual therapy, and general practitioner care for neck pain and showed that the manual therapy group improved faster than the physiotherapy and general practitioner care groups (Korthals-de Bos et al. 2003). The total costs of manual therapy (447€) were about one-third of the costs of physiotherapy (1297€) and general practitioner care (1379€).

Curing disease is no longer the only goal. Between biotech medicine and consumer product, between physiotherapy and sport, between medical device and video game, between curative treatment and cultural service, between hospital and hotel, a whole area of health solutions is emerging, with evidence-based psychological, physical, nutritional, digital, and elemental interventions, in short, NPIs.

The weak regulation of the NPI ecosystem constitutes an advantage for this promising macro market, a major domain of investment of academics, independent investors, and industries.

13 Financing Models

Let us try to list the potential funders of NPIs in the next 10 years, not necessarily in order of importance.

13.1 Public Health Insurance

Guided by the World Health Assembly resolution WHA64.9 from May 2011 and based on the recommendations from the World Health Report 2010 "Health systems financing: The path to universal coverage," WHO is supporting countries in the development of health financing systems that can bring them closer to universal coverage. The financing levers to move closer to universal health coverage rely on three interrelated domains: raising funds for health, reducing financial barriers to access through prepayment and subsequent pooling of funds in preference to direct out-of-pocket payments, and allocating or using funds in a way that promotes efficiency and equity. Developments in these key financing domains will determine whether health services exist and are available for everyone and whether people can afford to use health services when they need them. In France, public health insurance reimburses some NPIs directly. For example, acupuncture provided by a doctor is reimbursed at the rate of approximately 55 € per session. A phytotherapist consultation is reimbursed at around 25 € per session. Some specific initiatives are permitted with a "cancer care package" and with the article 51 "Law public health 2018 – Organizational innovations for the transformation of the health system". The provision of wigs for cancer have started to be reimbursed by health insurance during cancer treatments since 2019. Some financial supports come from welfare part of health insurance.

Public health insurance indirectly reimburses NPIs via their global allocations to hospitals. Department heads and directors might, in some cases, cut their budgets and their daily fees to offer yoga, Tai Chi, Pilates, and hypnosis sessions during anesthesia. Some hospitals receive donations and through this can use NPIs without impacting their operating budget.

The increasing focus on the costs of care is forcing healthcare organizations to look critically at their basic set of processes and activities, to determine what type of value they can deliver. A business model describes the resources, processes, and cost assumptions that an organization makes that will lead to the delivery of a unique value proposition to a customer. As healthcare organizations are beginning to transform their structures in preparation for a value-based delivery system, understanding business model theory can help in the redesign process.

The National Health Insurance Fund for salaried workers has been experimenting in three French departments (Haute-Garonne, Bouches-du-Rhône, and Morbihan) since 2017 with the reimbursement of an NPI for people aged 18–60 years suffering from depression of mild to moderate intensity: offering a psychotherapy. The experiment is based on collaborative work between doctors and psychologists on the basis of specifications that detail the course, the number and content of the sessions, the method of reimbursement (experimental treatment sheet), and the prices.

13.2   Private Health Insurance

Private health insurance companies are now interested in the issue of NPIs and are inspired by countries that are pioneers in this area, such as Switzerland, where about 60% of the adult population have a supplemental insurance that covers part of the complementary medicine treatments. The French health insurance AG2R launched a national NPI initiative in 2019 against recurrent cancer including a case management program involving several NPIs after biomedical treatment for cancer (*Branchez-Vous Santé* program).

Private Reimbursement, Not for All Conditions

After conducting pilot experiments and retrospective analyses, private health insurance companies realized that a blind reimbursement for "alternative medicines" (where the patient alone chose the healthcare professional and the practice) did not work. The use of digital systems has enabled them to move to a second phase, that in which reimbursement of NPIs is conditional on real participation by patients and compliance with specifications followed by approved professionals.

An insurer offers its policyholders the opportunity to accumulate points in order to obtain rewards from partners after providing proof of regular physical activity by a connected object. Another insurer offers NPI checks from 7000 steps made daily.

In Germany, an individual pension plan commits the insured to lead a healthier life by rewarding them with a reduction in their insurance premiums and reduction vouchers with approved partners (e.g., gym, approved organic supermarket).

American companies impose wellness programs on their employees if they do not want to lose their health insurance. These programs use gamification and personalization strategies to facilitate the lasting engagement of members.

These economic models bet on lower expenditure in terms of biomedical treatments but also better health and productivity of employees.

Businesses have understood that human capital is essential and that it must be preserved. Through procedures imposed by law for the prevention of psychosocial or voluntary risks, companies (driven in particular by competition) offer specific NPIs in relation to stress management, dietary advice, prevention of low back pain, and smoking cessation. Of course, these strategies are indirectly designed to prevent absenteeism and attract the best workers.

13.3 Crowdfunding

Financing models for NPIs can also be based on fundraising through non-profit national foundations and local associations. Donations can be received. Special events can be organized by a national foundation or local association to raise funds (e.g., racing, sporting event, show, exhibition).

Dragon Ladies

The Canadian doctor, Dr. Ronald McKenzie, physiotherapist and oncologist, formed the first crew of Dragon Ladies in 1996. Opposing the medical recommendations of the time, he demonstrated that the practice of an adapted physical activity, such as rowing, is recommended to prevent lymphedema in all women who have undergone axillary dissection and mastectomy. The first crews paddling dragon boats were born in the United States, then Australia and in Europe. Today, the official recommendations of the Cancer Plan recommend the practice of regular adapted physical activity because it is now proven that it considerably reduces the risk of recurrence, up to 40% for breast cancer.

13.4   Public Investment

Public institutions can invest a part of their budget in activities aimed at improving human health through funds intended for social and/or environmental activities. Private-public partnerships can be proposed, for example, with a free loan of materials or premises and with the temporary provision of personnel.

13.5   Direct Taxes

High salt intake, coming mainly from processed foods, contributes to high blood pressure, which in turn increases the risk of stroke, chronic kidney disease, coronary heart disease, and diabetes. Regular intake of some other foods such as those high in fat, artificial ingredients, hormones, antibiotics, and genetically modified organisms (GMOs) also present the risk of premature death. Alcohol, cigarettes, and marijuana are dangerous, especially for cancer. These products are subject to taxation differently in different countries. The money collected from these taxes could be used for the dissemination of relevant NPIs.

13.6   Tax Deduction

The state can also adjust taxes as a function of governmental strategy. France, for example, reduces taxes (e.g., VAT) from 20% to 5% for owners who renovate their accommodation with more energy-efficient and environmentally friendly features and for restaurateurs who cook better-quality dishes. The state may allow people to deduct specific fees from their taxes.

Support for strengthening social solidarity in the economy can be provided at state level such as, for example, assisting startups in the health field, which means the state can benefit from a total exemption from salary and employer charges during the first 2 years of operation.

13.7 Territorial Administrations

Regions, districts, and cities work together to enhance their territory and make themselves visible, in some cases beyond the state. The activities of the population and the region's employment capacity represent points of major attractiveness for a territory. Aware of the potential of activities that the NPI sector represents in connection with health and wellness, these administrations have been creating clusters in tourist places, in areas close to innovation campuses (e.g., university of technology, medical school, silver economy business valley, cosmeceuticals industries zone), in residential areas for the elderly (e.g., healthy-aging communities), and in hospital areas (e.g., healthcare business center).

13.8 Health Data, as Money

Some companies want to help people monetize their own health and wellness information, beyond the ethical concerns of civil liberties advocates. Selling personal health data is completely legal in the United States, for instance. If the data belongs to the person until their death and is conceived as a new natural resource, a copy can be sold. Anybody could request a copy of their electronic health record from doctors and post it on eBay or another electronic marketplace. The amount of data being produced is large, from biomedical (e.g., lab test results, blood pressure readings, DNA determinations, illnesses, therapeutics inventories, surgeries, medical imaging, vaccinations, allergies) to subjective experiences (e.g., personal rates, experience feedback, survey answers, preferences) to socio-demographic information (e.g., gender, age, profession, marital status, income), to profiling of behaviors (e.g., nutrition, leisure, smoking habits, alcohol consumption), up to billing history. For example, a U.S. firm, Hu-manity.co, attempts to connect users of its app to companies and organizations using a "consent as a service" model. They extract medical information from users who "consent for privacy" and "authorize for permitted use." Corporations give free services to consumers in exchange for monetizing their data. Thus, if consumers do not want their data monetized, they cannot have free services. These companies are actually seeking the U.S. government's help in creating a market for it. Politicians look into "data dividend" strategies for consumers as a payout from Big Tech. These companies monetize data every single second; why not be rewarded for it? In January 2019, 40 lawmakers in Oregon signed onto a bipartisan bill that would enable consumers to monetize their de-anonymized medical data (Oregon State Legislature 2019).

13.9   Research and Innovation

States are facing serious healthcare challenges with each citizen hoping for a long, happy, and healthy life (Chap. 2). Aging and the expansion of chronic diseases require innovative solutions and organizations to maintain care costs at a sustainable level and to limit the risk of unequal access to care. In addition, external environmental factors such as climate change, or public health issues such as inappropriate behavior and fragility (e.g., antimicrobial resistance, infectious diseases), are also exposing people to new risks and threats. States, supranational organizations (e.g., the European Union), and various foundations are investing in research, technology, and innovation to develop solutions to overcome these challenges. The return on this investment will be finding new frameworks to prevent diseases, developing better diagnostics, discovering more effective therapies, as well as taking up new models of care and new technologies promoting health and wellness. Innovations could keep older people active and independent for longer and help care systems to remain sustainable. The U.S. National Center for Complementary and Integrative Health has invested US$2.5 billion since 1998 for research and information on NPIs. Europe is starting to take an interest in it. Australia, China, France, Germany, India, Japan, and Switzerland actively support researchers with specific grants, even if they are not yet sufficiently visible and detached from conventional unmarked research funds.

13.10   User Contribution

Finally, NPI users are not sheep who are forced into imposed solutions that are supposed to be good for their health. This vertical and paternalistic healthcare model belongs to the nineteenth century, when the family doctor did what he could, or to the twentieth century, when an omniscient hospital doctor-in-chief decided for the patient. Many people already contribute, without even knowing it, to the financial participation in taking care of their health: by consuming food supplements, visiting their osteopath, using mHealth instruments.

The twenty-first century is building actors responsible for their health and wellness and who wish to collaborate with qualified multidisciplinary teams for shared decision making. As such, to take a share of the responsibility is to commit financially. According to the calculation base to be decided by the regulator, contributing to this health effort does not seem unimaginable.

14   Conclusion

The popularity of NPIs is growing; the supply of NPIs is diversifying; the market is igniting. NPIs are accessible everywhere, from hospitals to community-based medicine, from pharmacies to e-commerce, from hotels to shopping centers. A vast and promising market has been launched since 2010. This represents a boon for professional corporations and for personal care and services: a limitless area of innovation. The world market for NPIs was estimated at US$359 billion in 2017. It is increasing by 5% each year.

Usage affects all generations, ages, sexes, sociocultural levels, and countries. The demand concerns wellness, self-knowledge, healthy aging, disease prevention, biomedical treatment potentiation, and other forms of care in the name of freedom of choice. The potential of NPIs goes beyond individual and collective health care. The field interests the tourism, digital, food, and distribution industries. This weakly regulated market is full of opportunities for wellness service professionals; for entrepreneurs, manufacturers, investors, insurers, and bankers; and for the media. The NPI market is not a bubble because it integrates all of the various sources of current changes: epidemiological-demographic transition, digitalization, globalization, rationalization, and segmentation (personalization, pluralism of choice) in a field that has no equal to bring people together, and that is health. It is a component of a future of health defined by radically interoperable data, open yet secure platforms, and consumer-driven care (Allen 2019, p. 32).

Key Points

- The last study dating from 2017 places the world market for NPIs at US$359 billion; within the global wellness market estimated at US$4220 billion. It increases by 5% each year.
- About 66% of people use NPIs. The use attracts all generations, ages, sexes, sociocultural levels, and countries.
- The leading countries in the sector are China, India, Germany, Japan, Switzerland, and the United States.
- The demand encompasses wellness, self-knowledge, healthy aging, disease prevention, the potentiation of biomedical treatments, and other forms of care.
- The NPI field, going beyond individual and collective health care, attracts interest from tourism, food, digital, and distribution industries.
- The weakly regulated market is a boon for caregivers, wellness professionals, delivery services, manufacturers, investors, insurers, bankers, and media actors.
- The NPI market is not a bubble because it integrates all of the sources of current changes: epidemiological-demographic transition, digitalization, globalization, rationalization, and segmentation (personalization, pluralism of choice) in a field that has no equal for bringing people together – health.

References

Allen, S. (2019). *2020 global healthcare outlook. Laying a foundation for the future*. Canberra: Deloitte.

Atout France. (2015). www.atout-france.fr

Begot, A. C. (2008). Médecines parallèles et cancers: Pratiques thérapeutiques et significations sociales. *Revue Internationale sur le Médicament, 2*(1), 50–95.

CAMbrella. (2012). *A pan-European research network for complementary and alternative medicine*. Brussels: CAMbrella.

Center for Strategic Analysis. (2012). Quelle réponse des pouvoirs publics à l'engouement pour les médecines non conventionnelles? *Note d'Analyse, 290*, 1–11.

Centre de Recherche et d'Information des Organisations de Consommateurs. (2012). *Les Médecines alternatives*. Brussels: CRIOC.

Di Sarsina, P. R. (2017). The social demand for a medicine focused on the person: The contribution of CAM to healthcare and healthgenesis. *Evidence-based Complementary and Alternative Medicine, 4*(S1), 45–51. https://doi.org/10.1093/ecam/nem094.

Dilhuydy, J. M. (2003). L'attrait pour les médecines complémentaires et alternatives en cancérologie: une réalité que les médecins ne peuvent ni ignorer, ni réfuter. *Bulletin du Cancer, 90*(7), 623–628.

Eisenberg, D. M., Kessler, R. C., van Rompay, M. I., Kaptchuk, T. J., Wilkey, S. A., Appel, S., et al. (2001). Perceptions about complementary therapies relative to conventional therapies among adults who use both: results from a national survey. *Annals of Internal Medicine, 135*(5), 344–351. https://doi.org/10.7326/0003-4819-135-5-200109040-00011.

European Chronic Disease Alliance. (2019). *Towards an EU strategic framework for the prevention of non-communicable Diseases (NCDs)*. Brussels: ECDA.

French National Authority for Health. (2011). *Développement de la prescription de thérapeutiques non médicamenteuses validées*. Paris: HAS.

Global Wellness Institute. (2017). *Global wellness economy monitor*. Miami: GWI.

Global Wellness Institute. (2018). *Global wellness economy monitor*. Miami: GWI.

Graz, B., Rodondi, P. Y., & Bovin, E. (2011). Existe-t-il des données scientifiques sur l'efficacité clinique des médecines complémentaires? *Forum Médical Suisse, 11*, 808–813.

iCEPS conference. (2020) www.icepsconference.fr

IFOP. (2007). *Les Français et les médecines naturelles: résultats détaillés*. Paris: IFOP.

Jonas, W. B., & Levin, J. S. (1999). *Essentials of complementary and alternative medicine*. New York: Lippincott Williams and Wilkins.

Khalil, M. K. M., Al-Eidi, S., Al-Qaed, M., & AlSanad, S. (2018). The future of integrative health and medicine in Saudi Arabia. *Integrative Medicine Research, 7*, 316–321. https://doi.org/10.1016/j.imr.2018.06.004.

Klein, S. D., Torchetti, L., Frei-Erb, M., & Wolf, U. (2015). Usage of complementary medicine in Switzerland: Results of the Swiss health survey 2012 and development since 2007. *PLoS One, 10*(10), e0141985. https://doi.org/10.1371/journal.pone.0141985.

Korthals-de Bos, I. B., Hoving, J. L., van Tulder, M. W., Rutten-van Mölken, M. P., Adèr, H. J., de Vet, H. C., Koes, B. W., Vondeling, H., & Bouter, L. M. (2003). Cost effectiveness of physiotherapy, manual therapy, and general practitioner care for neck pain: economic evaluation alongside a randomised controlled trial. *British Medical Journal, 326*(7395), 911–916. https://doi.org/10.1136/bmj.326.7395.911.

Mao, J. J. (2017). Advancing the global impact of integrative oncology. *Journal of the National Cancer Institute Monographs, 52*, 1–2. https://doi.org/10.1093/jncimonographs/lgx001.

Marguerit, D., & Reynaudi, M. (2015). Quelle place pour la France sur le marché international des soins? *Note d'Analyse, 27*, 1–8.

Martin, K. I., & Glaser, D. A. (2011). Cosmeceuticals: The new medicine of beauty. *Missouri Medicine, 108*(1), 60–63.

Molassiotis, A., Fernandez-Ortega, P., Pud, G., et al. (2005). Use of complementary and alternative medicine in cancer patients: A European survey. *Annals of Oncology, 16*, 655–663. https://doi.org/10.1093/annonc/mdi110.

Morin, L., & Laroche, M. L. (2016). Utilisation des médicaments chez les personnes de 60 ans et plus en France: un état des lieux à partir des données de l'Assurance maladie. *Revue de Gériatrie, 41*(6), 335–349.

OECD. (2015). *Health Panorama de la santé 2015: les indicateurs de l'OCDE.* Paris: OECD Editions.

OECD. (2016). *Panorama de la santé 2015: comment se positionne la France?* Paris: OECD Editions.

Oregon State Legislature. (2019). https://olis.leg.state.or.us/liz/2019R1/Measures/Overview/SB703

Pandey, A., Jatana, G. K., & Sonthalia, S. (2020). *Cosmeceuticals.* Treasure Island: StatPearls Publishing.

Simon, L., Prebay, D., Beretz, A., Bagot, J. L., Lobstein, A., Rubinstein, I., & Schraub, S. (2007). Médecines complémentaires et alternatives suivies par les patients cancéreux en France. *Bulletin du Cancer, 94*(5), 483–488.

Smith, T., Gillespie, M., Eckl, V., Knepper, J., & Morton-Reynolds, C. (2019). Herbal supplement sales in US increase by 9.4% in 2018. *HerbalGram, 123*, 62–73.

Society for Integrative Oncology. (2020) www.integrativeonc.org

Thomas, K., Nicholl, J., & Fall, M. (2001). Use and expenditure on complementary medicine in England: A population-based survey. *Complementary Therapies in Medicine, 9*, 2–11.

WHO. (2001). *Legal status of traditional medicine and complementary/alternative medicine: A worldwide review.* Geneva: WHO.

WHO. (2002). *Herbal WHO traditional medicine strategy 2002–2005.* Geneva: WHO.

WHO. (2013). *WHO traditional medicine Strategy: 2014–2023.* Geneva: WHO.

WHO. (2019). *WHO global report on traditional and complementary medicine 2019.* Geneva: WHO.

Willcox, M. L., & Bodeker, G. (2004). Traditional herbal medicines for malaria. *British Medical Journal, 329*, 1156–1159. https://doi.org/10.1136/bmj.329.7475.1156.

Chapter 8
Evaluation of Non-pharmacological Interventions

1 Introduction

Non-pharmacological interventions (NPIs) radically divide opinion. There are pros and cons as much in the ranks of clinicians as among researchers, patients, and decision makers. The skeptics want to understand everything before using them and recall the many falsifications that have undermined confidence in NPIs throughout the history of medicine. At best they are no more effective than placebo solutions. Those who are convinced of the merits of NPIs stress the fact that they work with people around them and that they have observed benefits for themselves, a close person, or their patients. As mentioned by Edzard Ernst (2009, p. 299), *"testing the efficacy and safety of CAM is not just an obsession of scientists; it is a concern for the health of individuals and the public at large."* This chapter presents the scientific challenges that are being overcome in the course of the last 10 years.

2 Demand for Effectiveness

In a world dominated by rationality and single-causality, explaining that a mechanism of action is at work in a "complementary medicine empirically tinkered with" is no longer enough (Chap. 3). Doubts and uncertainty must be avoided in the health field. The patient who becomes a client demands a guaranteed benefit with minimum risk and expense. He or she wants to avoid the mistakes of the past where decisions were made blindly or intuitively.

Overcoming Controversies: The Story Is Not New

Since many people believe in the practice of alternative medicine, it is tempting to assume that it is safe. *"According to this argumentum ad populum, millions cannot all be wrong. However, belief can be wrong, practice can be misguided, and popularity is not a reliable indicator for effectiveness; after all, medicine is no popularity contest. The history of medicine is littered with examples that demonstrate how misleading this fallacy can be."* (Ernst 2013, p. 1025). For example, between 1795 and 1796, the American doctor Elisha Perkins invented and patented wands with healing powers, known as the *"Perkins Tractors."* Made from an original metal alloy, they seemed to relieve various illnesses such as rheumatism and headache. They had to be passed along the nerves of the body to the areas of inflammation. The English epidemiologist John Haygarth replicated these treatments in similar patients using Perkins chopsticks or wooden chopsticks. He obtained the same results with the two types of chopsticks, four out of five patients declaring themselves to be much better. John Haygarth thus experimentally exposed the hoax, which he published in 1800 in a work entitled "On the curious influence of the imagination on the functions of the human body." He then highlighted two pillars of science and clinical research: reproducibility and statistical superiority. If a therapeutic solution only works once, then you cannot establish a causal link. At the same time, if an innovative treatment does not have a statistically superior benefit to the conventional treatment, then it does not have to be used.

Authorities and health funders, and most importantly patients, are requiring evidence of effectiveness and safety of NPIs, specifications for implementation, and trained and responsible professionals (Fig. 8.1).

Demand for the evaluation of the effectiveness of NPIs is accelerating under the pressure of six categories of actors:

- Patients who have become organized and informed consumers demanding to know the real effects of NPIs, the risks incurred for health, and the constraints of use
- Health professionals wishing to offer their patients practices based on scientific evidence to differentiate themselves from charlatans
- Researchers, convinced today by mechanistic studies and observational cohorts (Behrens et al. 2013)
- Academic societies encouraging the construction of consensual prescription decision trees and good practices
- Health insurance companies and provident organizations wishing to clarify the responsibilities of each, estimate the risks in the event of a problem, and reimburse the NPI at the best value for money
- Decision makers faced with an epidemiological transition unprecedented in human history (exponential increase in the number of elderly and people with chronic disease) who are waiting to know which are the best NPIs before taking action

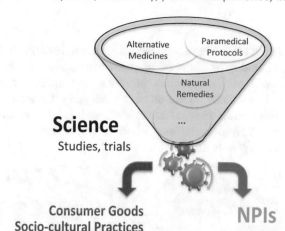

Opinion, testimony, personal experience, intuition

Fig. 8.1 Demand for effectiveness of NPIs

Steve Jobs's Lesson

In October 2003, doctors for Apple co-founder and CEO Steve Jobs announced that he had pancreatic cancer. They recommended surgery to remove the tumor. He refused, and for 9 months turned to alternative medicines. The media, and his biography published 19 days after his death (Isaacson 2011), discussed a vegetarian diet, acupuncture, and plants (Isaacson 2011). A new examination revealed that the tumor had progressed. Surgery performed on July 31, 2004 removed the head from the pancreas, gallbladder, part of the intestine, and stomach. This major operation probably indicated that the cancer had progressed and that metastases had started to spread. A further aggravation led him to try an experimental treatment of radiotherapy, without success. He died on October 5, 2011, at the age of 56, of a very aggressive cancer, an endocrine tumor of the islets of Langerhans. Endless controversies ensued, some accusing natural medicines, others experimental radiotherapy, others again the imposition of a liver transplant and anti-rejection drugs. Do you think he wanted to die? Do you think that a man so brilliant and wealthy would have made a decision going against the state of science at the time? Obviously not. Scientific data was lacking to enable him to make the best decision. This example shows that even the wealthiest, most intelligent person, with the greatest number of medical resources, cannot make informed decisions about the use of NPIs or biotechnology therapies if the information does not exist. (Greenlee and Ernst 2012).

3 Medical and Institutional Caution Contrasts with Evolving Practices

Numerous published pilot studies have been reporting benefits of NPIs in improving health, autonomy, and quality of life in participants with disease or frailty. These studies can sometimes signal a reduction in direct and indirect health costs. Unfortunately, these observations do not prove it irrefutably. Collective expertise published by authorities remain reserved due to methodological shortcomings. For example, the latest French Nutrition Agency report did not recommend fasting during cancer treatments and post-treatments (ANSES 2010).

> **Recurring Doubts of the Authorities**
> *With regard to the criteria usually considered for evaluating the effectiveness of drug treatments, studies evaluating the effectiveness of non-pharmacological therapies present for the most part methodological shortcomings.* (French National Authority for Health 2011, p. 40).

Without recommendations from the authorities, NPIs will remain the subject of endless contradictory debate, provocatively illustrated by a simplistic position evident in an editorial title by Edzard Ernst "*alternative practitioners amuse the patient, while medics cure the disease*" (Ernst 2018). They will continue to attract individuals fond of paranormal explanations such as crabs to heal cancer (Atshan et al. 2020), questionable testimonies, improbable amalgams, manipulations of all kinds, Perkins magic wand tricks, etc. Their chances of systematic integration into the care pathways and reimbursement by insurance authorities will be compromised in a sector subject to strong economic, political, and corporate lobbies.

4 Drugs Framework

It was not that long ago, only around 50 years in the past, when drugs were at the same level of uncertainty as NPIs today (Bouvenot and Vray 2006). It was clinical and experimental research that helped remove doubts and wipe out dangerous practices. It is interesting to remember that the first clinical trial in the world was a non-pharmacological solution, Dr. James Lind's famous citrus scurvy diet (Chap. 2).

> **50 Years Ago, Drugs Were Also Subject to Doubts**
> "Until the 1960s, many therapeutic interventions [drugs] had as their only justification, so to speak, the force of routine, the credulous attachment to traditions, or a generalization from a few occasional and anecdotal examples improperly called professional experience" (Bouvenot and Vray 2006, p. 13). Everything changed with the adoption of a single, standardized model for drug validation and monitoring.

Demonstrating the effectiveness of an NPI on health means providing irrefutable evidence via a clinical trial of its effect on one or more endpoints based on a sufficient number of representative participants and a finding of significant statistical difference between the group testing the NPI and the control group – in other words, demonstrating significant differences, and not just observing appearances. An NPI will be demonstrated to be effective if its administration to people in an experimental group shows a greater benefit than that evident in a control group taking a placebo or following routine care.

Such an experiment takes place in the context of a randomized controlled trial. This procedure is highly supervised and monitored. Statistically and clinically, there must therefore be a significant difference between the experimental group and the control group at the end of the intervention to conclude that the NPI is effective. In other words, a statistically significant difference in a group of patients over time is insufficient to demonstrate the effectiveness of an NPI. In endorsing this method, this book advocates that NPIs follow the course of evidence-based medicine (EBM).

5 Evidence-Based Medicine, the Only Future for NPIs

Following the logic of the methodology involved in evidence-based medicine is to offer science-based content to the solution of health problems. The benefits and risks of each NPI should be known. Validated NPIs should be proposed after a reliable checkup and available diagnosis and offered at an optimal time. Any proposition should respect the user's preferences and obtain their agreement. It is a question of offering NPIs in complete safety, aiming:

– To prevent the occurrence, worsening, and complication or recurrence of a disease (prevention)
– To improve health status and increase lifespan with a good quality of life (care)
– In some rare cases, to cure an illness (cure)

If part of the effect will always fall on the singularity of the relationship between a patient and a healthcare professional, the "rest" will fall within the framework of practice and the intervention method itself. This "rest" can be described, observed, filmed, studied, and compared as such (Hoffmann et al. 2014).

Over time, the effect related to the singular experience of the patient in the relationship with the healthcare professional ends up explaining the variance in the benefits obtained, including in areas where the quality of the therapeutic alliance is particularly important, such as psychotherapy (Horvath et al. 2011; Haug et al. 2012). The charisma of the healthcare professional and the quality of the therapeutic alliance are necessary but insufficient conditions for the success of an NPI. Studies indicate that a psychotherapeutic method counts more in the medium term than the person who performs it (Horvath et al. 2011; Haug et al. 2012).

Misunderstandings About Evidence-Based Medicine (EBM)

The EBM approach was developed by epidemiologists at McMaster University in Canada in the early 1980s, in particular by Professor David L. Sackett. It aims to base preventive and therapeutic decisions on scientific evidence. Sackett and colleagues caution that without clinical expertise "*practice could fall under the tyranny of evidence*" since even the most excellent experimental evidence may be inapplicable or inappropriate for a patient. This point is not always understood. The concept of evidence corresponds to knowledge obtained from clinical research having obtained valid results and that is applicable in current clinical practice. Clinical studies considered are randomized controlled trials, meta-analyses, cross-sectional studies, cohort studies, and well-constructed case-control studies (Sackett et al. 2000). Academic societies and advisory commissions are increasingly using this framework to issue a recommendation for an intervention targeting a particular population. These "proofs" do not replace judgment and experience; they complement them. They also help healthcare and prevention professionals to follow recent innovations, improve their practice, and, in turn, perfect and further specialize each NPI. The French National Authority for Health has recommended the use of the EBM grid since 2013 in France (HAS 2013).

The EBM approach is based on three pillars: the rigorous, explicit, and judicious use of the best current evidence in decision making for the individual treatment of patients; the need to include patients' choices in the decision-making process; and clinical expertise.

The knowledge of the effectiveness of a health solution is classified according to the level of evidence, from best to least good:

- Level A: meta-analysis based on data from randomized controlled trials with a good confidence index and follow-up of more than 80%
- Level B: systematic review of cohort studies, individual cohort study, or randomized controlled trial with follow-up of less than 80%
- Level C: systematic review of case-control studies, quality case-control studies
- Level D: case series, cohort or case-control study
- Level E: expert opinion or exploratory study

The Advantages of EBM

1. Updating of knowledge in fields that are progressing rapidly on a global scale
2. Comparison of the effectiveness of the interventions
3. Increased team confidence in decision making
4. Standardization of good interventional practices
5. Rationalization of resources following a quality approach

6. Better understanding of human behavior
7. More critical analysis of the results of clinical studies
8. More precise targeting of the objectives to be achieved
9. Better training of practitioners
10. Facilitation of intra-team, inter-team, patient, and family communication
11. Continuous improvement of each intervention

Any NPI should, therefore, be able to demonstrate that its benefit does not result from a placebo effect or the natural course of a health disorder depending on the lifestyle of the individual (Chap. 2). This recent development opens the door to an immense field of innovation, to the rediscovery of some old forgotten methods, and to the arrival of practices little known in the West. The application of EBM to evidence-based practice (EBP) is conceived, according to Straus et al. (2019), as the integration of the best available research evidence with clinical expertise and with the patient's unique values and preferences (i.e., personal concerns, expectations, cultural influences, and individual characteristics during the clinical encounter).

6 Pre-Market Research

6.1 Supports

Despite the complexity and immensity of the work that this represents, many agencies have supported research and innovation in the field of NPI development since the early 2000s, including the National Institutes of Health (NIH) in the United States, the National Institute for Care and Health Excellence (NICE) in England, and the Department of Health in Australia.

> **A Question of Motivation**
> In 1986, a renowned French Professor of Medicine, Daniel Schwartz, wrote: *"the rigorous evaluation of complementary medicines is sometimes impossible, most often difficult. But many evaluations, total or partial, much more than is recognized, remain possible. Their realization requires above all that we want it. Most of the time we don't want it"* (Schwartz 1986, p. 87). This desire has existed in the NPI field for 10 years.

In Europe, there is no agency dedicated to the systematic application of EBM to NPIs. This is despite the conclusions of the European research program CAMBrella that were published in an online report in 2012, the mobilization of professionals in the sector, and the repeated demand from user associations. Scattered initiatives are emerging, different from one country to another, and not always lasting.

In France, the state currently leaves the field open to suitable structures such as CUMIC, GETCOP, OMNC, and Plateforme CEPS as well as to interested academic research units and hospitals.

A Global Mobilization

French agencies such as the National Agency for Research on AIDS and Viral Hepatitis, the Institute for Research in Public Health, the National Cancer Institute, and the National Alliance for Life Sciences and Health (AVIESAN) created a working group to improve the transferability of intervention research. The French Academy of Medicine invites practitioners and researchers to base prevention on a true culture of science (2013). Patient associations, research foundations, and learned societies actively contribute to the conduct of clinical studies. Hospitals are opening up to non-pharmacological clinical trials with the support of hospital clinical research programs, universities, and public laboratories. Private clinics, spas, medical homes, and health networks are also contributing to this national effort.

6.2   Increase in Pragmatic Trials

The non-pharmacological scientific and medical literature shows evidence of an increase in pragmatic trials, such as those recommended by Professor Daniel Schwartz, for example (Schwartz and Lellouch 2009). These intervention studies demonstrate the effectiveness of NPIs, alone, in combination with, or in addition to biological treatments (Chap. 4).

The rigor of the NPI protocols has been improving (Boutron et al. 2008, 2012). The science of solutions complements the science of problems working on the mechanisms involved (basic research) (Hawe and Potvin 2009). In other words, the pragmatic study of the conditions for implementing NPIs has become as important as understanding the principles of action. Specialists speak of intervention research as that aiming to show the effectiveness (real-life effect) and not efficacy (effect of highly selective or overly restrictive participant selection in an artificial laboratory context) of an innovative health solution.

6.3   Improvements in the Methodological Quality of Studies

Since 2010, the number of controlled trials verifying the efficacy of NPIs has increased despite the financial, human, and material resources required by these protocols. They are gaining statistical power, increasing the number of subjects included, and improving outcome measure instruments. They are gaining in methodological rigor (internal validity) and in transferability to real life (external validity). Systematic reviews of intervention studies are a key for motivation, especially in areas that are less easy than testing health products (e.g., plants, food supplements, cosmetic products) such as education programs. For example, a systematic review of 85 controlled studies in educational interventions encouraged the use of high-quality instruments to measure outcomes (Albarqouni et al. 2018). The evaluation criteria are broader: benefits (effectiveness on health determinants and health-related quality of life), risks (side effects and risks of interaction with other therapies), utility (cost-effectiveness), and constraints (additional burden, non-adherence).

The description of the NPIs evaluated is more precise and more exhaustive, although there is still room for improvement when compared to that for drugs (Glasziou et al. 2008). Comparison groups are more credible.

A collaborative study conducted within an academic structure is a guarantee of rigor, transparency, ethics, and a limitation of conflicts of interest. Randomization is also an important methodological aspect to scrutinize. For example, the standard of a double-blind trial required for testing drugs is often impossible in the NPI context. How can you hide from the patient and the healthcare provider that the healthcare provider is pricking the patient with acupuncture needles?

International standardization recommendations are emerging for the design of protocols for evaluating NPIs (Boutron et al. 2008; Chan et al. 2013; Hoffmann et al. 2014) with some essential specifications when compared to drugs or medical devices (Falissard 2015):

- Systematic registration of the protocol before its implementation (see, for example, the WHO with the ICTRP, the United States with ClinicalTrials, Europe with Clinical Trials Register, China with Chinese Clinical Trial Register, Brazil with Brazilian Clinical Trials Registry, Germany with German Clinical Trials Register, Iran with the Iranian Registry of Clinical Trials, Japan with Japan Primary Registries Network, and the Cochrane Central Register of Controlled Trials)
- Ethics (e.g., Committee for the Protection of Persons in France)
- Transparency of interest links
- Description of the NPI (e.g., addendum available online)
- Implementation techniques by professionals (Michie et al. 2013)
- Potential accessibility of data and protocol notebooks (e.g., Open Science)
- Systematic publication and follow-up of citations (Boutron et al. 2008)

The WHO Plan 2014–2023 for traditional and complementary medicines encourages innovation in evaluation methods and outcome criteria for NPIs given their specificity, in particular, because of their human mediation (WHO 2013). The WHO points, in particular, to the need for medico-economic studies (e.g., cost-effectiveness) likely to convince decision makers to recognize these practices and better reimburse them.

Mortality Is Not the Only Criterion

Although epidemiological evidence has shown that being physically active can provide a protective effect against cancer recurrence, cancer-specific mortality, and all-cause mortality for some types of cancer, no evidence to date has evaluated whether the effects of exercise can positively affect cancer survival—although there are several clinical trials underway assessing exercise as a treatment intervention. Indeed, despite having what seems like an insufficiently robust evidence base to inform a strong move in either direction, it is hugely encouraging to see that treatment interventions that focus on a patient's quality of life and wellbeing are being seriously incorporated into the patient experience, and that exercise is now considered a vitally important component of cancer care at both ends of the care pathway—both as a treatment and as a preventive measure. (Lancet Oncology Editorial 2018, p. 715).

WHO also encourages the implementation of real-world experiments combining several methods and several scientific disciplines to consolidate the available knowledge and evidence-based practices.

For Real-World Experiments

While there is much to be learned from controlled clinical trials, other evaluation methods are also valuable. These include outcome and effectiveness studies, as well as comparative effectiveness research, patterns of use, and other qualitative methods. There is an opportunity to take advantage of, and sponsor such "real world experiments" where different research designs and methods are important, valuable and applicable. (WHO 2013, p. 39).

Many experts encourage research to go further in this direction (Boutron et al. 2012; Ninot 2013, 2019). They invite researchers to use mixed methods combining qualitative and quantitative techniques. The challenge is just as much about verifying the statistical superiorities of groups testing an NPI, as it is about understanding the participants' experience (Gueguen et al. 2014). Last but not least, studies need to assess possible side effects because any therapy involves risks (Falissard 2016).

7   Post-Market Research and Surveillance

7.1   Users

The evaluation of NPIs is not just a matter of researchers, manufacturers, and creators. Users are now invited to report benefits, interactions, side effects, and abuses. Platforms are created for the good of the community. Consumers can score their experience, register appreciation of what worked and indicate what didn't work, and testify to drug interactions, usage constraints, experiences, and failures. They must become whistle-blowers, if necessary, directly to health authorities or indirectly via social networks.

7.2 Practitioners

The data collected on NPIs comes from practitioners. They can rate the best and worst NPIs to resolve a health condition. Moreover, healthcare professionals can also assess NPI practitioners on specific digital platforms (e.g., *Medoucine* in France).

> **Transparency and Honesty**
> *Complementary and alternative medicine practitioners must be competent and adequately educated; their practice must be based upon good evidence; patients must be accorded autonomy; everyone involved in promoting or practicing complementary and alternative medicine must behave honestly; and patients and consumer of complementary and alternative medicine must never be exploited.* (Ernst and Smith 2018, p. xxv).

7.3 Authorities

The market surveillance system for safety of herbal medicines is relatively close to that for drugs, and in many countries benefits from drug monitoring procedures (WHO 2019). Morocco monitors herbal medicines through the *Centre Anti-Poison et de Pharmacovigilance Du Maroc* [Poison Control and Pharmacovigilance Centre of Morocco]. Pakistan has an official government website to report reactions to herbal medicines. Malaysia subjects herbal medicines to similar criteria that have been established for drugs, that is, regulation, surveillance, pharmacovigilance, licensing, and reporting of adverse drug reactions.

Some countries have created specific entities for following the safety and benefits of NPIs such as National Center for Complementary and Integrative Health (NCCIH) in the United States or the University College of Integrative and Complementary Medicine (CUMIC) in France (Chap. 1).

However, most countries currently use conventional health surveillance systems to carry out the post-market evaluation of NPIs, unfortunately without making NPIs a priority as they are presumed to be less potentially dangerous than biomedical treatments.

8 Decoding the Results of Intervention Studies

A reading of any single study published on an NPI in a scientific journal should question the context of the study, and, therefore, its relevance as a robust evaluation. More and more clinical studies are performed effectively "in-house" by research laboratories, with regard to the sums invested and the time pressures to publicize a product. Journalists sometimes read a study too quickly and superficially, or neglect methodological limits and the reservations expressed by researchers. Others can be challenged by the analyses carried out. Others may focus on a positive outcome of the study and ignore the negative results, side effects, and constraints for patients. Others overemphasize a sub-analysis for which the protocol was not originally designed (in the case of drug studies, for example, where patients could choose an NPI). Others jump too quickly to conclusions that confirm their prior opinions. Others wanting to be concise and educational generalize the effectiveness of an NPI to other health problems.

The speed at which information spreads is so spectacular with social media (Chap. 2), it is almost viral. It is easier to read an interpretation of the original article and relay it than to read the original article in-depth and synthesize it yourself. Let us not even mention partial or full plagiarism and conflicts of interest.

Misinterpretation of Statistics

In 2012, Franz Messerli published the results of a promising study for the treatment of Alzheimer's disease in the world's most famous scientific and medical journal, the *New England Journal of Medicine*. The study is based on human cohorts showing that a subclass of flavonoids, called flavanols, can slow and even reverse the decline in cognitive performance that occurs with age. The consumption of flavanols, very present in cocoa, was associated with better results in neuropsychological tests in patients with mild neurodegenerative disease. One study showed improved cognitive performance in rats after administration of polyphenolic cocoa extract. Biologists have suggested that flavanols improve endothelial function and lower blood pressure, causing vasodilation in the peripheral vascular system and the brain. Messerli's study formally establishes a linear, strong, and statistically significant ($r = 0.79$, $p < 0001$) correlation between per capita chocolate consumption and the number of Nobel Prize winners per 10 million people in 23 countries (Messerli 2012). The regression line allows Messerli to recommend an annual rise of 400 g of chocolate per capita to increase the number of Nobel prizes in a country by one each year. A minimum effective dose is 2 kg per year, with no maximum limit. The author thus confirms a dose-dependent relationship between the consumption of chocolate and the improvement of cognitive functions. This work alerts us to errors in the interpretation of statistics. Even a strong correlation is a relationship of proportionality, not causation. It indicates that either X influences Y, or Y influences X, or X and Y are influenced by a common underlying mechanism. This work raises awareness of the problems in making short cuts between animal studies and studies following human cohorts in order to apply them to everyday life. We are all tempted to reach Messerli's conclusion, without understanding the methodology and its limits, no offense to the chocolate industry. Many clinical studies remain to be done to prove the therapeutic efficacy of NPIs, either alone or in addition to other treatments.

Scientific journals are uneven in accepting articles on NPIs. Some no longer charge readers to access an article, but charge researchers to publish an article which distorts the quality of the expertise. New scientific communication vectors have appeared over the last 10 years (e.g., on a platform such as *Arxiv*, an Open science system such as *HAL*, or a preprint server for Health Sciences such as *medRxiv*). Biased studies are still too often published in the field (Ioannidis 2014). Confirmation bias may play a role in the clinical translation of new evidence from clinical trials (Tilburt et al. 2010).

Systematic reviews list all of the intervention studies evaluating the effectiveness of an NPI in solving a health problem (e.g., Cochrane Reviews). The fact that they are systematic means that their authors have made every effort to include all the studies and not to select those that were suitable for their methodology. It is, therefore, necessary to verify that an exhaustive selection procedure has been conducted and ensure it has included all the trials and justified all exclusions. It is also important to be careful in terms of efficacy criteria (e.g., pain measured with the same instrument). It is also necessary to identify the characteristics of the NPI. Many systematic reviews lack sufficient details about the interventions, and thus fail to be conclusive and able to be reproduced (Hoffmann et al. 2017). Too vague a criterion (e.g., depression) or too broad an NPI description (e.g., dietary supplement) will result in the review having little value.

9 Decoding Opinions of Authorities

Health authorities are publishing reports on NPIs whose findings remain inconclusive, even contradictory between agencies, and often call for further studies. They underscore the complexity of the mechanisms brought into play by the NPIs in question (Chap. 3). They point to insufficiencies in the description of NPIs and the lack of additional adequate description for comparison and for implementation. A study showed that only 34% of non-pharmacological clinical trials offer additional information on the Internet (Hoffmann et al. 2013). Health authorities also criticize the heterogeneity of assessment protocols.

At the same time, it is their role to monitor closely the health of all. We must, therefore, be patient and see how much these institutions are also progressing in understanding a sector as dynamic as that of NPIs.

10 A Need for a Consensual Validation and Surveillance Framework

Real progress has been made over the last decade to make the evaluation of NPIs more rigorous. The number of randomized controlled trials has increased exponentially across the five NPI categories. Nothing will replace them as long as science relies on the Popperian approach, requiring scientific method to be based on the "formulation of refutable hypotheses by reproducible experiments" (Falissard 2005, p. 5). These experiences are the only bulwark against drifts of all kinds, especially in disciplines focusing on some aspect of human life (Macleod et al. 2014). They should remain the keystone in demonstrating the effectiveness of NPIs in target populations (Boutron et al. 2012). Minimal methodological requirements in a unique validation and surveillance framework should be required and known to all, in order to serve meta-analyses and better justify the usefulness of new clinical trials. This standard has been the case for drug development for 50 years. Researchers and manufacturers share a consensual framework which guides them from the laboratory to authorization and implementation post-marketing. According to the guidance of the U.S. Food and Drug Administration (FDA), the framework characterizes the collection and evaluation of data, from the process design stage through commercial production, in which scientific evidence of efficacy, safety, and quality of product are consistently established. The principles and methods are similar across Western countries. The framework is organized in five phases: phase 0 (preclinical) to identify mechanisms, phase 1 to determine tolerance in healthy humans, phase 2 to identify the optimal dose for a small number of patients (pilot trial), phase 3 to demonstrate evidence of efficacy and safety (randomized controlled trial), and phase 4 to ensure long-term safety. A narrative review that sorted and categorized validation and surveillance frameworks on NPIs has been published in the scientific and medical literature (Carbonnel and Ninot 2019). The results showed the absence of a unique and consensual model (Fig. 8.2). In April 2019, 46 frameworks coexist, without one of them being predominant or even showing a convergence emerging toward one of them. Their numbers have increased exponentially over time since the 1970s. The United States and United Kingdom are the main countries to propose frameworks, probably due to their university research productivity and new technology industry leadership.

In the same way for drugs 50 years ago, it is urgent to adopt a consensual framework for validation and monitoring of NPIs (Carbonnel and Ninot 2019). If, naturally, there will always remain an element of singularity due to the contribution of the relationship between the practitioner and the patient, it becomes urgent to adopt on a supranational, even international scale, a consensual approach of verification of the safety and effectiveness of NPIs through rigorous intervention studies, optimization of practices through implementation studies, and usage monitoring analyses

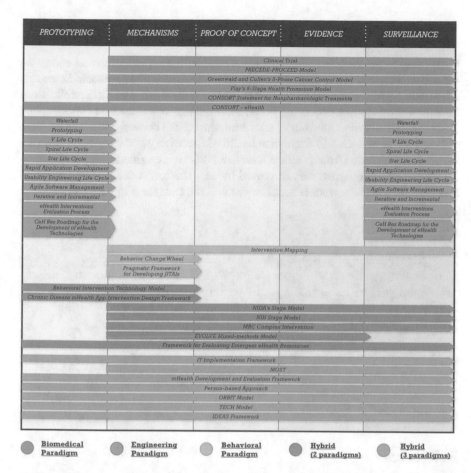

Fig. 8.2 Validation and surveillance frameworks for NPIs published in the scientific and medical literature. Reprinted from Carbonnel, F., Ninot, G. (2019). Identifying Frameworks for Validation and Monitoring of Consensual Behavioral Intervention Technologies: Narrative Review. Journal of Medical Internet Research, 21(10), e13606. Figure 8.1. doi: https://doi.org/10.2196/13606, licensed under the terms of the Creative Commons Attribution License (https://creativecommons. org/licenses/by/4.0/). ©François Carbonnel, Gregory Ninot. Originally published in the Journal of Medical Internet Research (https://www.jmir.org/2019/10/e13606), 16.10.2019

using big data and artificial intelligence. For example, a study published in the journal *Nature* showed that major non-pharmaceutical interventions have had a large effect on reducing transmission of COVID-19 in Europe (Flaxman et al. 2020). Such an approach would undoubtedly help to remove the barriers, skepticism, and stubborn false beliefs regarding NPIs. Big data analyses and patient opinions will count in this process.

11   Conclusion

Conversely to practitioners of alternative medicine, those producing and endorsing the use of NPIs resolutely follow the evidence-based medicine approach. Science is advancing slowly, counterintuitively, but surely. With about 2 million publications of intervention studies evaluating NPIs, their number has increased exponentially since 2000. Study after study, trial after trial, the results converge to show the benefits and limits of each NPI for human health. Unnecessary and dangerous NPIs are isolated and excluded from practice. Relevant NPIs are described with precise specifications and are progressively improved by new studies. A consensual validation and surveillance framework is necessary as was the case for drugs 50 years ago.

Key Points

- NPIs resolutely follow the approach of evidence-based medicine.
- About 2 million publications of intervention studies evaluating the effectiveness of NPIs, representing an exponential increase since 2000.
- Results converge to show the benefits and limits of each NPI for human health through systematic reviews.
- Unnecessary and dangerous NPIs are isolated and excluded.
- Relevant NPIs are described with precise specifications.
- A supranational, even international, consensual framework of validation and surveillance is needed on NPIs as was the case for drugs 50 years ago.
- Greater involvement of patients will be requested in the evaluation of NPIs.

References

Albarqouni, L., Hoffmann, T., & Glasziou, P. (2018). Evidence-based practice educational intervention studies: A systematic review of what is taught and how it is measured. *BMC Medical Education, 18*(1), 177. https://doi.org/10.1186/s12909-018-1284-1.

ANSES. (2010). *Évaluation des risques liés aux pratiques alimentaires d'amaigrissement*. Paris: ANSES.

Atshan, S. S., Abduljaleel, S. A., & Alessa, H. S. (2020). Should crab heals cancer humans? *Traditional Medicine Research Cancer, 3*(1), 28–31. https://doi.org/10.12032/TMRC201800065.

Behrens, G., Fischer, B., Kohler, S., Park, Y., Hollenbeck, A. R., & Leitzmann, M. (2013). Healthy lifestyle behaviors and decreased risk of mortality in a large prospective study of U.S. women and men. *European Journal of Epidemiology, 28*(5), 361–372. https://doi.org/10.1007/s10654-013-9796-9.

Boutron, I., Moher, D., Altman, D. G., Schulz, K. F., & Ravaud, P. (2008). Extending the CONSORT statement to randomized trials of nonpharmacologic treatment: Explanation and elaboration. *Annals of Internal Medicine, 148*, 295–309. https://doi.org/10.7326/0003-4819-148-4-200802190-00008.

Boutron, I., Ravaud, P., & Moher, D. (2012). *Randomized clinical trials of non pharmacological treatments*. Bacon Raton: CRC Press Taylor and Francis.

Bouvenot, G., & Vray, M. (2006). *Essais cliniques: Théorie, pratique et critique*. Paris: Lavoisier.

Carbonnel, F., & Ninot, G. (2019). Identifying frameworks for validation and monitoring of consensual Behavioral intervention technologies: Narrative review. *Journal of Medical Internet Research, 21*(10), e13606. https://doi.org/10.2196/13606.

Chan, A. W., Tetzlaff, J. M., Altman, D. G., et al. (2013). SPIRIT 2013 statement: Defining standard protocol items for clinical trials. *Annals of Internal Medicine, 158*(3), 200–207. https://doi.org/10.7326/0003-4819-158-3-201302050-00583.

Ernst, E. (2009). Ethics of complementary medicine: Practical issues. *British Journal of General Practice, 59*(564), 517–569. https://doi.org/10.3399/bjgp09X453404.

Ernst, E. (2013). Thirteen follies and fallacies about alternative medicine. *European Molecular Biology Organization, 4*(12), 1025–1026. https://doi.org/10.1038/embor.2013.174.

Ernst, E. (2018). Alternative practitioners amuse the patient, while medics cure the disease. *Journal of Clinical Medicine, 7*(6), 137. https://doi.org/10.3390/jcm7060137.

Ernst, E., & Smith, K. (2018). *More harm than good? The moral maze of complementary and alternative medicine*. Cham: Springer.

Falissard, B. (2005). *Comprendre et utiliser les statistiques dans les sciences de la vie*. Paris: Masson.

Falissard, B. (2015). How should we evaluate non-pharmacological treatments in child and adolescent psychiatry? *European Child and Adolescent Psychiatry, 24*(9), 1011–1013. https://doi.org/10.1007/s00787-015-0762-9.

Falissard, B. (2016). Les "médecines complémentaires" à l'épreuve de la science. *Recherche et Santé, 146*, 6–7.

Flaxman, S., Mishra, S., Gandy, A., et al. (2020). Estimating the effects of non-pharmaceutical interventions on COVID-19 in Europe. *Nature*, https://doi.org/10.1038/s41586-020-2405-7.

French Academy of Medicine. (2013). *Thérapies complémentaires – acupuncture, hypnose, ostéopathie, tai-chi – leur place parmi les ressources de soins*. Paris: French Academy of Medicine.

French National Authority for Health. (2011). *Développement de la prescription de thérapeutiques non médicamenteuses validées*. Paris: HAS.

French National Authority for Health. (2013). *Niveau de preuve et gradation des recommandations de bonne pratique*. Paris: HAS.

Glasziou, P., Meats, E., Heneghan, C., & Shepperd, S. (2008). What is missing from descriptions of treatment in trials and reviews? *British Medical Journal, 336*(7659), 1472–1474. https://doi.org/10.1136/bmj.39590.732037.47.

Greenlee, H., & Ernst, E. (2012). What can we learn from Steve jobs about complementary and alternative therapies? *Preventive Medicine, 54*(1), 3–4. https://doi.org/10.1016/j.ypmed.2011.12.014.

Gueguen, J., Hill, C., & Barry, C. (2014). Complementary medicines. In *Wiley StatsRef: Statistics reference online*. Hoboken: John Wiley & Sons, Ltd. https://doi.org/10.1002/9781118445112.stat05556.pub2.

Haug, T., Nordgreen, T., Öst, L. G., & Havik, O. E. (2012). Self-help treatment of anxiety disorders: A meta-analysis and meta-regression of effects and potential moderators. *Clinical Psychology Review, 32*, 425–445. https://doi.org/10.1016/j.cpr.2012.04.002.

Hawe, P., & Potvin, L. (2009). What is population health intervention research? *Canadian Journal of Public Health, 100*, I8–I14.

Hoffmann, T. C., Erueti, C., & Glasziou, P. P. (2013). Poor description of non-pharmacological interventions: Analysis of consecutive sample of randomised trials. *British Medical Journal, 347*, f3755. https://doi.org/10.1136/bmj.f3755.

Hoffmann, T. C., Glasziou, P. P., Boutron, I., Milne, R., Perera, R., Moher, D., Altman, D. G., Barbour, V., Macdonald, H., Johnston, M., Lamb, S. E., Dixon-Woods, M., McCulloch, P., Wyatt, J. C., Chan, A. W., & Michie, S. (2014). Better reporting of interventions: Template for intervention description and replication (TIDieR) checklist and guide. *British Medical Journal, 348*(g1687), 1–10. https://doi.org/10.1136/bmj.g1687.

Hoffmann, T. C., Oxman, A. D., Ioannidis, J. P., Moher, D., Lasserson, T. J., Tovey, D. I., Stein, K., Sutcliffe, K., Ravaud, P., Altman, D. G., Perera, R., & Glasziou, P. (2017). Enhancing the usability of systematic reviews by improving the consideration and description of interventions. *British Medical Journal, 358*, j2998. https://doi.org/10.1136/bmj.j2998.

Horvath, A. O., Del Re, A. C., Fluckiger, C., & Symonds, D. (2011). Alliance in individual psychotherapy. *Psychotherapy, 48*(1), 9–16. https://doi.org/10.1037/a0022186.

Ioannidis, J. P. (2014). How to make more published research true. *PLoS Medicine, 11*(10), e1001747. https://doi.org/10.1371/journal.pmed.1001747.

Isaacson, W. (2011). *Steve Jobs*. New York: Simon & Schuster.

Lancet Oncology Editorial. (2018). Exercise and cancer treatment: Balancing patient needs. *Lancet Oncology, 19*, 715. https://doi.org/10.1016/S1470-2045(18)30376-0.

Macleod, M. R., Michie, S., Roberts, I., Roberts, I., Dirnagl, U., Chalmers, I., Ioannidis, J. P. A., Al-Shahi Salman, R., Chan, A. W., & Glasziou, P. (2014). Biomedical research: Increasing value, reducing waste. *Lancet, 383*(9912), 101–104. https://doi.org/10.1016/S0140-6736(13)62329-6.

Messerli, F. H. (2012). Chocolate consumption, cognitive function, and Nobel laureates. *New England Journal of Medicine, 367*(16), 1562–1564. https://doi.org/10.1056/NEJMon1211064.

Michie, S., Richardson, M., Johnston, M., Abraham, C., Francis, J., Hardeman, W., Eccles, M. P., Cane, J., & Wood, C. E. (2013). The behavior change technique taxonomy (v1) of 93 hierarchically clustered techniques: Building an international consensus for the reporting of behavior change interventions. *Annals of Behavior Medicine, 46*(1), 81–95. https://doi.org/10.1007/s12160-013-9486-6.

Ninot, G. (2013). *Démontrer l'efficacité des interventions non médicamenteuses: Question de points de vue*. Montpellier: Presses Universitaires de la Méditerranée.

Ninot, G. (2019). *Guide professionnel des interventions non médicamenteuses (INM)*. Paris: Dunod.

Sackett, D. L., Strauss, S. E., Richardson, W. S., & Haynes, R. B. (2000). *Evidence-based medicine: How to practice and teach EBM*. New York: Churchill Livingstone.

Schwartz, D. (1986). Peut-on évaluer les médecines douces? *Sciences Sociales et Santé, 4*(2), 75–88. https://doi.org/10.3406/sosan.1986.1034.

Schwartz, D., & Lellouch, J. (2009). Explanatory and pragmatic attitudes in therapeutical trials. *Journal of Clinical Epidemiology, 62*(5), 499–505. https://doi.org/10.1016/j.jclinepi.2009.01.012.

Straus, S., Glasziou, P., Richardson, W. S., & Hayne, S. B. (2019). *Evidence-based medicine: How to practice and teach EBM* (5th ed.). China: Elsevier.

Tilburt, J. C., Miller, F. G., Jenkins, S., Kaptchuk, T. J., Clarridge, B., Bolcic-Jankovic, D., Emanuel, E. J., & Curlin, F. A. (2010). Factors that influence practitioners' interpretations of evidence from alternative medicine trials: A factorial vignette experiment embedded in a national survey. *Medical Care, 48*(4), 341–348. https://doi.org/10.1097/mlr.0b013e3181ca3ee2.

WHO. (2013). *WHO traditional medicine strategy: 2014–2023*. Geneva: WHO.

WHO. (2019). *WHO global report on traditional and complementary medicine 2019*. Geneva: WHO.

Chapter 9
The Future of Non-pharmacological Interventions

1 Introduction

Despite the growing success of non-pharmacological interventions (NPIs) in different countries that were highlighted in the previous chapters, it is important to consider their weak points that must be addressed and resolved in a transparent manner in the coming years. This chapter presents the challenges concerning organization, research, monitoring, regulation, training, and the prospects for the resolution of these issues.

2 Organization

2.1 A Common Language

Beyond this book's clarification of NPI terminology, there is an urgent need to use the concept of NPIs within the frame of evidence-based health methods in every country.

The term "*alternative medicine*" and equivalents such as "alternative therapies," "natural medicine," "complementary and alternative medicine," "traditional medicine," and "integrative medicine" engender confusion about the reality of what NPI practices are and who are their qualified practitioners. This distinction is essential because these terms mix diagnosis and health interventions into a single category. As a result, the professionals of these unconventional medicines practice without control, supervision, and prerequisite healthcare training (Chap. 1). "*Some alternative therapies claim to treat or cure cancer, but while some people may say they are helpful, there is no scientific evidence that they have any benefit*" (Irish Cancer Society 2018, p. 10). NPIs are methods with verified benefits on health determinants and traceability that should not be equated to the *alternative medicines* that attempt to avoid assessment and surveillance, at the risk and peril of each user.

The term *"supportive care"* and its equivalents (e.g., "comfort medicine," "complementary therapy") are also not relevant to quality NPIs, which cannot be discounted, made secondary, or restricted to entertainment. The effective delivery of NPIs at the proper time within a personalized health strategy is part of a coherent whole.

The term *"public health message"* is also too vague and broad to be used in reference to NPIs, which cannot be restricted to principles of general lifestyle. The implementation of NPIs needs to follow protocols that carefully combine dose, content, and ingredients in order to ensure their sustainable and significant effects. Furthermore, their use must be formally known and agreed upon by both the user and a specifically trained professional.

Going further, the term *"complex intervention"* (Craig et al. 2008) is also too confusing, even if the use of NPIs simulates multimodal and systemic mechanisms that affect multiple determinants that mediate human health. This term amalgamates various exercise methods with psychotherapeutic protocols, mobile health devices, and herbal remedies into a single nomination.

As such, the use of the term *"NPIs"* specifically defined as methods of evidence-based practice methods guarantees both an optimal level of protection as well as the consideration, on the part of authorities, of false advertisements and gurus (Fig. 9.1). Evidence-based practice integrates the best available research evidence with clinical expertise and the unique values and preferences of patients such as personal concerns, expectations, cultural influences, and individual characteristics (Straus et al. 2019).

Fig. 9.1 An NPI conceived as a manualized and evidence-based method for human health

The need for the direct delivery of personalized interventions to patients places NPIs in a unique position that is full of responsibility and promise. Given the diversity of NPIs and the types of health conditions they address, it is important that the term carries a comprehensive definition that adheres to science and industry-adopted core principles and best practices (Chap. 1).

2.2   A Transparent Inventory Duty

There are about 11,000 drugs classified in official registers, but how many NPIs exist – 1000? 10,000? 100,000? Many reports confuse systems with approaches, mechanisms (also called processes), methods (also called protocols or programs), components (also called active ingredients or elements), procedures (also called techniques), and materials (e.g., fins and snorkel for swimming training). Some translations of NPIs in various languages are ambiguous, leading to drama for patients such as in the case of Chinese herbal remedies (Chap. 5). NPIs are medicines in their own right insofar as their effectiveness has been proven and their risks identified. Many reports reduce the benefits of NPI as a product of a good patient-practitioner relationship, even though NPIs do not use the same mechanisms of action (Chap. 3).

Several health authorities, such as the French National Authority for Health (2011), are calling for better information on the use of NPIs. An exhaustive list of these practices is currently missing. This could take the form of a multilingual repository of NPI designations and full descriptions that reduce confusion about their definition and practice (Table 1). This inventory would also allow patients to be informed about the names of the methods that are proposed by a professional, recommended by a relative, or seen on a website.

Knowing the name of an NPI is necessary but insufficient; a listing of the specifications relevant to each NPI should be created despite the tedious nature. One of the characteristics of the NPI concept that was emphasized throughout this book calls for adequate descriptions of these practices so that they can be evaluated, monitored, and improved (Table 9.1).

This would provide NPI users the ability to refer to leaflets, user's manuals, and studies published in scientific journals that assess the health benefits and risks of any given practice. This kind of clear labeling is being called for by scientists who are encouraging authors to accurately and completely describe NPIs (Boutron et al. 2012; Carbonnel and Ninot 2019; Hoffmann et al. 2013). These descriptions should include things like session-by-session overviews for education programs or component-by-component guides to food supplements. Imagine how many accidents have resulted from poorly practiced manual therapies, and how many misunderstandings with patients that could have been avoided if these reference materials were in place. Too many therapists are still guided by nothing other than their intuition or current mood.

Table 9.1 Expected categorized characteristics of each NPI.

	Obligatory	Optional
Designation	Name	Acronym(s), synonym(s), author(s), institution(s), label(s)
Health goal	Main health problem that it prevents, treats or cures	Secondary benefits
Target population	Minimum–maximum age	Gender, ideal socio-educational level, territory
Content	Components (active ingredient, or gesture), procedures (session, dose, duration), materials	Professional and user manuals
Context	Place of practice, optimal delivery time within the health pathway	The requirements for medical prescription, reimbursement
Mechanisms	Verified mechanisms	Hypothetical mechanisms
Provider	Provider profession	Initial training diploma, continuing education diploma, qualification, certification
Scientific publications	≥ 1 publication of a positive intervention study	Additional publications (other methods, systematic reviews, collective expertise)

What Everybody Should Know Before Trying an NPI

- Its scientific name
- Its main health objectives
- Its secondary health benefits
- Its risks and precaution-prevention strategies
- Its ideal patient population
- Its content, especially its active ingredients
- Its mechanisms of action
- The diplomas, qualifications, and certifications of its practicing professionals
- The scientific publications and authority reports that recommend it (Table 9.1)

Considering this logic, the need to create a global database of effective NPIs that brings together these types of standardized specifications and their respective scientific publications becomes clear. For the time being, the research on NPIs consists of scattered and uneven studies that do not describe these practices homogeneously. It appears as though it is time to build an evolving library that is useful to users and practitioners alike. With the number of publications on NPIs increasing worldwide, the use of information systems and artificial intelligence is essential to this endeavor, as it would allow for the construction of a multilingual classification that is based on the contributions of both experts and users. Not only would this facilitate access to quality studies, it would also save users from having to wait for the release of the new edition of Dr. Marc Micozzi's book every 4 years (Chap. 1).

2.3 Accessible Notices for All

The European Medicines Agency (EMA), the U.S. Food and Drug Administration (FDA), and other health agencies require drug firms and biomedical device manufacturers to provide standardized information for patients in the form of package leaflets that include information such as the drug's pharmacotherapeutic category, therapeutic indications, contraindications, precautions for use, potential side effects and risky interactions, and dosage. This should also be required for NPIs.

> **Information Notices for NPI Professionals and Individuals Are Needed**
> There is a clear need for a standardized notice pertaining to each NPI for professionals and users. The only persons who would be against manualized NPIs are charlatans.

Food supplement manufacturers are required to specify the composition of their products according to the established standards in the countries in which they are marketed. If NPIs continue to be considered as a collection of separate domains (e.g., herbal medicine, manual therapy, exercise program, psychotherapy, health application, serious games), users have no way of knowing if different NPIs can treat the same health condition in the same way or if various NPIs would be more effective if used in combination.

Pharmacists will only recommend the plants, mushrooms, and minerals that they know, and, in particular, that they sell the most of. Psychologists will only recommend the psychotherapy they pertain to and its ramifications. Physiotherapists will only mention the methods they have learned. Dietitians will only specify the diets they have mastered. Acupuncturists...

Creating and relying upon standardized information notices pertaining to specific NPIs would allow:

– Professionals to discuss, using a common framework, the scientifically established percentage of success and risks known by science with their patients and with other health professionals
– Patients to make an informed choice about their preferred NPIs, allowing them to make a lasting commitment with knowledge of the practice's benefits and risks

2.4   Transparent Professions

NPI-related professions must shed light on their practices. For example, do you know the difference between a nutritionist and a dietitian, between a psychologist and a psychotherapist, between a physiotherapist and an osteopath? More complicated, do you know the difference between a sophrologist and a health coach, between a chiropractor and a kinesiologist? Do you know what the training is for naturopaths or healers? Do you know that there are different schools of thought and practice in both reflexology and osteopathy? Do you know that there are no regulations in France for the creation of a psychoanalytic practice?

Doctors have observed that patients are becoming extremely demanding and sometimes procedural with their biomedical therapies. Yet, for the moment, it is curious to note how weak this level of requirement is for NPIs and their practitioners. Most patients do not bother to read the professional plaque hanging on the wall or to notice that their practitioners' titles, expertise, and diplomas are often eccentric or even illegal. Some corporations have no ethics codes and no national organization.

> **The Confederation of Practitioners of Natural Medicine in Quebec**
> The *Confédération des Praticiens en Médecine Douce du Québec* (CPMDQ 2020) is a Quebec-central syndicate that was created in 1991 in order to bring together practitioners in naturopathy, massage therapy, kinesiology, orthotherapy, homeopathy, kinesitherapy, sports training, hypnotherapy, acupressure, traditional Chinese medicine, and various forms of alternative medicine. The organization promotes the self-healing of the body through natural therapeutic techniques both in the public sphere and in the private sector with its many partners.

The fact that an NPI professional gains a reputation by word of mouth or media exposure is not enough to qualify them as a competent professional. This is why it is becoming urgent to provide patients with transparent information on diplomas, on qualifications acquired in continuing education, and on the exact missions of each NPI profession. For example, online and app-based directories and appointment scheduling systems are working to meet user demand, but they still face many difficulties in classifying NPI trades, approaches, methods, and techniques (e.g., Medoucine (2020) in France).

2.5   New Delivery Models

Western medicine has reduced itself to a practice in which one organic problem = one biotechnological solution (Chap. 2). As a result of this extreme simplifi-
cation, some doctors have acquired a Pavlovian reflex; for each health problem,
there is a unique biomedical treatment. Human health, however, is far more complex, involving many simultaneous interactions.

As such, the growing demand for NPIs acts as evidence of a renewed interest in
renewing the true ancestral role of the doctor as defined by Hippocrates – the one
who listens, the one who understands the patient as a whole, the one who establishes
the diagnosis, and the one who advises the best preventive and therapeutic solutions.
The doctor is also the one who prevents the patient from falling into the traps of
dangerous practices (*primum non nocere*). Taken together, it becomes clear that a
patient-centered approach and NPIs will be key points for twenty-first-century medicine (Oude Engberink et al. 2017).

NPIs Will Benefit from the Digital Transformation of the Health System
The development of personalized and predictive care pathways in moderniz-
ing preventive and participatory health system represents a real opportunity
for NPIs by allowing for:

- Better prescription
- Better association
- Better risk anticipation
- Better information

The digital transformation of health systems involves the emergence of a comprehensive and holistic approach that will boost the integration of NPIs. These technologies facilitate new forms of continuous, monitored, open, and secure data
platforms (e.g., wearable sensors, big data analytics); new delivery models (e.g.,
machine learning, automatic services); and new opportunities for collaboration
among all stakeholders. Cloud computing solutions are accelerating flexibility,
interoperability, and mobility. They are also reducing time-consuming tasks (e.g.,
artificial intelligence), giving access to high-documented applications (e.g., health
open data hub, health knowledge platform), and improving team communication
(e.g., 5G technology). The digital transformation is making healthcare information
more accessible by encouraging providers and insurers to share data with each other
and to improve the quality and efficiency of health care. For patients it is helping
them to make informed decisions and to exercise a more proactive adherence (e.g.,

natural language processing, augmented reality, 3-D holograms, flexible and inter-active learning systems, workgroup collaboration tools). New digital technologies are enabling healthcare organizations to share data, which can help people get the right NPIs at the right time, and in the right place, which improves both their safety and effectiveness.

2.6   A Label

This book provides principles aimed at helping practitioners and patients to make relevant decisions on the use of NPIs; but, this kind of time-consuming effort is limited and unsustainable in the medium term. Thus, a visible labeling system or repository is necessary as a means of guiding consumers in their choices of NPI solutions – whether they are in association with or in place of biomedical treatments and products. A label acts as a guarantee of stability and security in the NPI field. A national, continental, or international label provides guarantees of the efficiency and safety of each NPI by working to exclude dangerous and sectarian remedies that take advantage of regulatory ambiguity in order to abuse or exploit people experiencing health problems. NPI labeling will also define practices so that they can be advertised under the surveillance of appropriate authorities (e.g., written, printed, or graphic matter accompanying or associated with the NPI).

In this vein, a blockchain system that creates a digital record of transactions would be a useful way of helping healthcare organizations to monitor and track NPI presence and use.

2.7   A Transnational Agency

In 2016, Australia established a research center on complementary and integrative medicine at the University of Technology Sydney (UTS), the Australian Center in Complementary and Integrative Medicine (ARCCIM). Norway created Norway's National Research Center in Complementary and Alternative Medicine (NAFKAM 2020). The Stanford Faculty of Medicine instituted a Center for Integrative Medicine in 1998 and currently has 10 clinical trials underway. The Universities of Texas and of Maryland both have integrative medicine centers. The Academic Consortium for Integrative Medicine and Health is a global but mainly North American organization with members from 70 highly esteemed academic medical centers and health systems. The donor-supported Consortium was born in 2002 and holds annual conferences with titles such as "The Consortium on Integrative Medicine (2020)." Even the Cochrane collaboration (2020), a global independent network providing high-quality medical treatment evidence, has created a full-fledged specialty in complementary medicine. So, isn't it time to create a European or international agency on the subject of NPIs?

The National Center for Complementary and Integrative Health (NCCIH)

Established in 1998, the National Center for Complementary and Integrative Health (NCCIH 2020) is one of the 27 institutes of the National Institutes of Health that are attached to the U.S. Department of Health and Human Services. Its mission is to clarify, through rigorous scientific research, the usefulness and safety of NPIs. It also assists authorities in their roles in improving health and care. This national center relies on scientific data as a means of enlightening the general public, health professionals, and decision makers on the use of complementary and integrative approaches in health. It is committed to advancing basic science, to developing assessment methods, to improving care, to promoting health, and to preventing disease. In 2019, the Center had a budget of US\$146,473,000, representing a 9% increase from 2018. It has received \$2.5 billion USD since its initial inception. Its staff includes 68 full-time professionals.

2.8 An Innovative Ecosystem

As mentioned in Chap. 7, a vast and promising NPI ecosystem has been developing since 2010. This field gathers care and service professionals from various industries such as biomedicine, food, new technology, and tourism. Its users span all ages, sexes, socioeconomic levels, and countries.

The revered family doctor, the sacred hospital doctor prescribing prolonged immobilization, interminable hospitalizations, the dismal hospice, the candid and passive patient – all of these are now things of the past. Today, we treat with technical platforms and coordinated multidisciplinary teams, which give new duties to doctors and new rights to patients. Transparency, shared decision making, personalized care pathways quality, and improved home life have become the new watchwords. In parallel to this organizational revolution that has taken place over the last two decades, another more disruptive revolution has also come. Progress in computing, biotechnology, and nanotechnology are orienting the future of medicine around creating a new human species – the hybridization of man and machine with a lifespan lengthened and capacities increased tenfold. A growing part of the Western world is dreaming of making superhumans that are smarter, more mobile, faster, more "beautiful," more resistant, and living for 150 years. Transhumanism is intended to protect humans from pathogens, to cure them of disease, and to protect them from the effects of aging with the help of biocompatible prostheses, exoskeletons, and algorithms. Humans have gained 30 years of life expectancy between 1950 and 2000. If in only 50 years we've gained as much as in a millennium, biotechnologies can add 70 more years to our lives in the next 50 years. The American scientist, academic, and entrepreneur Raymond Kurzweil predicted that man would fuse with artificial intelligence by as early as 2045, which will allow him to increase his intelligence by a factor of a billion (Kurzweil 2005)! Moreover, stem cell implantation will renew our damaged cells as cells from our own bodies, grown in laboratories, are transplanted in us to replace our failing tissues. Medical nanotechnologies will directly release molecules on targeted pathogens, and surgeons will be robots of absolute precision. Irradiation systems will no longer exceed their target. Micromachines manufactured by 3D printers using biocompatible materials will replace the imperfect organs of donors. Embedded systems will help us think faster, react sooner, and avoid accidents. Prostheses will be directly controlled by our brains. Systems monitoring our actions will warn us in case of danger. Diseases will be predicted and treated in advance. The current debate on bioethics demonstrates the border between a derivation and a visionary innovation. What is unethical for one generation may become commonplace for the next; the history of medicine is full of examples. Ethics do not stop progress; even though democratic countries are supposed to protect their people by law and regulation, what is prohibited in one country is not necessarily prohibited in another. We've all heard about stem cell transplants in the spinal cord of paraplegic patients occurring in China when most countries refuse to allow this procedure due to the lack of sufficient evidence in animals and humans. We've all also heard about the storing of umbilical cords or wisdom teeth in Switzerland so that their stem cells can be used for therapeutic purposes 1 day. Borders no longer stop the treatments and anti-aging medicine of multinational industries. After the conquest of the West, the exploration of the deep seabed, the scaling of the highest peaks, the great polar expeditions, space travel and the moon landings, it is now time to accomplish man's wildest dream – immortality. There is a great marketing argument to be made in order to convince investors to enter the future-oriented market of consumer health and anti-aging products. We are now applying the mechanisms of biological degeneration to healthy humans as a means of constantly regenerating them (Clarke et al. 2010). The ultimate goal is to consider death as a simple disease that can be cured (de Grey and Rae 2007). Transhumanism seeks to prepare our transition to a post-human reality (Hayles 1999) in which humans are genetically modified cybernetic organisms that are the next step up for our species. Man will become equivalent to a god (Harari 2017).

This revolution will undoubtedly also crossover into the NPI ecosystem according to three user categories:

- The "radicals" refusing biotechnology solutions and artifact gadgets while denouncing the increase in social and environmental inequalities, clinging to natural solutions, and fighting against all means of giving immortality to some human beings
- The "pragmatists" using relevant NPI protocols in addition to biotechnologies
- The "beta-testers" experimenting NPIs as a means of gaining wellness, autonomy, and longevity

The NPI ecosystem will leverage opportunities from digital and biotechnological success and failure in order to aggregate consumer-driven communities by gathering individuals who are focused on wellness and operating largely outside of the medical system (Chap. 7). These geographic or virtual communities may be segmented (e.g., seniors, patients with mental condition), self-formed (e.g., meditation, diet, serious games), social (e.g., Pilates, tai chi, weight management), or natural (e.g., sports, schools, employers). They will inexorably promote NPIs in a better structured ecosystem that maintains its inherent flexibility.

3 Research

3.1 The World Expansion of Research

More and more academic researchers around the world are involved in NPI assessment using observational and mechanistic research, and increasingly, more interventional research (Chap. 2). This exponential increase of NPI publication contrasts with the small and moderately growing research field in complementary and alternative medicines (CAM), for example, in Scandinavia, with an average of 120 publications per year (Danell et al. 2019). Their ambition is to address the areas of ignorance, low knowledge, and contradiction that were underscored by Cohen-Mansfield in 2018: *"at present, non-pharmacological interventions are like "orphan drugs" in that there are no organizations with a commercial incentive to invest large sums in testing them"* (Cohen-Mansfield 2018, p. 283).

Optimistically, the last decade acts as a testament to unprecedented financial research support for NPI research on different national scales. For example, the 2018–2022 French Health Strategy *"supported the development and encourage the evaluation of non-pharmacological interventions"* (French Ministry of Health and Solidarity 2018, p. 48). The NCCIH and other public national agencies (e.g., NAFKAM 2020) also significantly supported researchers by increasing calls for grants (Chap. 8). States are increasingly noting the overrepresentation of clinical trial data coming from the United States, which is working to legitimately finance research that can meet the needs of its people. Trials are being conducted almost exclusively on North American soil, with people whose lifestyles are not representative of populations in other regions of the globe. Other states are becoming aware of this problem and are trying to improve investment in research, in particular, by pooling funds dedicated to grants.

Nishare: An Open Science Database of Relevant Academic Documents Dedicated to NPI Assessment

The Nishare (2020) system collects, organizes, and shares academic documents related to the evaluation of NPIs. It was created in 2019 by the Plateforme CEPS and is freely accessible on the Internet (Chap. 1). It compiles, sorts, and provides access to master's reports (or equivalent), thesis manuscripts (PhD or equivalent), unpublished study reports, and slideshows devoted to the assessment of NPIs. Documents in all languages are accepted as long as they have a title and a summary in English and they are in a shareable pdf format.

Moreover, new private players have joined forces with academic laboratories and departments such as in the case of contract research organizations (e.g., Mooven) and living labs (e.g., I2ML). These contribute to the implementation of digital technology in research that drives innovation, improves access to and the affordability of data, enhances quality, improves real-time monitoring, and reduces costs through efficient models. Some cloud platforms and analytical algorithms have created new ways of achieving higher quality and faster reporting for researchers.

> **Niri: The Global Directory of NPI Researchers and Research Organizations**
> Niri (2020) is the international directory of researchers, research units, and clinical research centers that have published at least one study evaluating an NPI in a scientific or medical journal. The system is freely accessible on the Internet and was designed in 2019 by the Plateforme CEPS (Chap. 1). It identifies and maps all of the public and private scientific actors in the global NPI ecosystem.

Private-public collaborations for NPI research are increasing as a result of government incentives promoting healthy aging and chronic disease prevention. Additionally, lobbies and nonprofit associations are raising funds for research on specific conditions (e.g., Alzheimer, addiction, cancer, heart disease, diabetes). These initiatives are establishing smart health communities and creating sustained research bases that are operating largely outside of the academic, hospital-centric system. They are also mobilizing reactive and adherent participants. Sharing these goals will optimize:

- Value-based care
- Fraud reduction, waste reduction, and the abuse of resources
- Real-time monitoring
- New product/service development

Today researchers are benefitting from collaborative systems, group management, the secure sharing of relevant data, and quality monitoring. These approaches facilitate the implementation of NPI studies. International congresses, such as the *iCEPS Conference* (2020), then bear witness to these large-scale collaborations.

> **iCEPS Conference: The International Scientific Congress on NPIs**
> Research indicates that NPIs are essential solutions for improving the health, independence, quality of life, and longevity of human beings. The international iCEPS Conference (2020) provides an opportunity for scientific data sharing on NPIs and for innovations in economic modeling, organization, and regulation. This Congress accelerates discussions on the societal impact of

(continued)

NPIs, their relevant evaluation methods, the mechanisms of action involved in their efficacy, strategies for their integration into individual pathways, and on good practices. The scientific program is managed by the academic and collaborative platform, the "Plateforme CEPS." The Congress offers annual plenary lectures, poster sessions, workshops, round tables, and open sessions to the public. Since its first edition in Montpellier in 2011, the Congress has welcomed more than 4000 people of 32 different nationalities.

3.2 The Improvement of Protocols

The principles of transparency, integrity, relevancy, and transferability have guided the conception of NPI studies since 2010. Ethics committees and research project funding selections are increasingly vigilant. Systematic reviews and meta-analyses have highlighted real improvements on the quality of methods as well as studies that provide sufficient levels of evidence, for example, in non-pharmacological treatments for common chronic pain conditions (Agency for Healthcare Research and Quality – Skelly et al. 2020).

Many NPI trialists are calling for more realistic clinical trials (Chap. 2) that are related to "real life," such as pragmatic intervention studies showing their effectiveness (Schwartz and Lellouch 2009). High-quality requirements in the design of clinical trials have become gold standards (Chap. 8). Some experts have adapted the rigor that is commonly seen in drug trials in the behavioral sciences; for example, in the cases of exercise programs, diets, disease management programs, and psychotherapies (Powell et al. 2020). This adaptation is promising trends and opportunities for the future of NPIs, especially for the testing of manualized treatments aimed at improving behavioral, social, psychosocial, or environmental risk factors for chronic diseases (e.g., obesity, sedentary behavior, adherence to treatment, psychosocial stress).

As Anne Louise Germaine de Staël-Holstein (1766–1817), also known as Madame de Stael, once noted, the *"search for the truth is the noblest occupation of man; its publication is a duty"* (Stevens 1881). A key element in the future of NPIs will be high quality reporting that improves their related conduct, usability, and interpretation of systematic reviews (Hoffmann et al. 2015). For example, improvements have been made to improve trial protocols by avoiding redundancy and improving transparency. Metasearch engines, such as Motrial, are able to match relevant intervention study publications that assess the benefits and risks of NPIs, as well as their prior protocol declarations (e.g., Motrial 2020).

**Motrial, an Open Science Metasearch Engine for Interventional Studies
Assessing the Benefits and Risks of NPIs**
Motrial is a metasearch engine that is accessible on the Internet. It was created
in 2018 by the Plateforme CEPS (Chap. 1). This digital system facilitates the
identification of interventional studies on prevention or therapies from around
the world (e.g., before-after studies, controlled trials, randomized controlled
trials). It compiles and sorts publications from multiple databases (medical
sciences, biological sciences, psychological sciences, educational sciences,
social sciences). It also isolates main publications from derived publications
by linking them to their prior protocol declarations made to competent authorities. Once the information is made available, either automatically or through
the addition of the authors, the system provides details on the sponsors, ethics
committees, the country of origin, and the financial support for each study.
Motrial helps researchers learn about the state of NPI interventional research
so that they can design more relevant studies and conduct more comprehensive literature reviews.

3.3   The Patient Experience Record

Decisions on the therapeutic and preventive use of NPIs still relies too heavily on
health professionals, especially doctors, in particular. The generalization of digital
health data and new decision-sharing obligations on the part of authorities for the
benefit of patients are reducing the knowledge asymmetry between patients and
healthcare professionals. Qualitative studies on patients' lived experiences are
increasingly preponderant in this evolution. For example, the integration of self-
assessments, such as patient-reported outcomes (PRO), are raising the importance
of patient opinion and feelings in their healthcare decisions; they are becoming the
master of their own destinies. This patient-centered, integrative care approach is
becoming reinforced through research and the emergence of smart health communities with a consumer-driven vision.

3.4   The Improvement of Data Analysis

Many authors of systematic reviews discuss the difficulty of obtaining exhaustive
publications of studies in the NPI field. This is because the major health and medical
databases have specific orientations, including disciplinary priorities (e.g., database

in nutrition therapy or psychotherapy), method preferences (e.g., database exclusively made up of clinical trials), and cultural positions (e.g., Chinese databases). Cloud platforms and intelligent data analysis systems facilitate access to relevant studies for researchers, practitioners, decision makers, and users.

An Example with Kalya Research

In 2020, the French-American startup Kalya, which defines itself as the first actionable intelligence dedicated to NPIs, has launched Kalya Research (2020). The digital system is a virtual research assistant based on artificial intelligence that finds, sorts, and analyzes current worldwide scientific publications on NPIs. Many scientific databases do not cover the entire NPI field. PubMed, for example, is powered by the National Center for Biotechnology Information (NCBI); as such, it is focused on biotechnology and biotherapy. PsychInfo is dedicated to the psychological research corpus. Thus, many NPIs for prevention, human services, and various disciplines at the frontier of medicine or psychology with direct and short-term impacts on human health do not fall within the scope of traditional databases (e.g., nutrition, plant sciences, sports sciences, public health, education sciences, neurosciences, new technologies, economics). The Kalya Research system helps researchers and practitioners by providing them with a more exhaustive and real-time vision of the scientific literature dedicated to NPIs.

For example, of the 11,835 publications of relevant intervention studies assessing an NPI in breast cancer prevention or care (July 25, 2020), illustrated in the Fig. 9.2, Kalya Research (2020) found a majority of research from the United States (Fig. 9.3), using predominantly exercise, diet, and health education programs (Fig. 9.4) targeted to improve health-related quality of life (Fig. 9.5).

3.5 A Common Validation and Surveillance Framework

The classical and consensual framework for drug development, validation, and surveillance is not fully adapted to the characteristics of NPIs and their delivery modes, especially in terms of the recurrent requirement for double-blind randomized trials (Chap. 8). Researchers have proposed 46 alternative frameworks for NPIs (Carbonnel and Ninot 2019); yet, one of these frameworks has not been accepted by authorities as the gold standard.

The Entropy of Assessment Frameworks
Models that are inspired by the drug paradigm, such as the CONSORT model (Boutron et al. 2008, 2012), maintain the randomized controlled trial as the heart of their processes. Other models rely on theories of behavior change, such as the American ORBIT model (Czajkowski et al. 2015; Powell et al. 2020). Others are inspired by engineering, which is based on iterative procedures that improve the quality of health solutions such as the Agile model (Beck et al. 2001). Others offer hybrid models, such as the MOST model (Collins et al. 2007) based on three successive adjustment phases or the Medical Research Council framework for developing and evaluating complex interventions. These frameworks offer more or less restrictive solutions for upstream research before marketing (mechanistic and interventional), and downstream research that follows implementation (monitoring). These involve validation times, certainty levels, and diametrically different resources (human, material, and financial). To date, no model has been chosen as the standard by one or more authorities. The number of these models is increasing, and the paradigmatic heterogeneity that results does not facilitate the work of researchers, meta-analysis specialists, or even of the engineers and designers of NPIs. Instead they generate sterile debates between people who do not speak on the same level of assessment (Carbonnel and Ninot 2019).

Whatever model is adopted in the future, it should involve – as much as possible – observational, mechanistic, and interventional studies. These are necessary to properly observe, explain, and compare NPIs (Fig. 9.6). Indeed, observational stud-

Results by Year

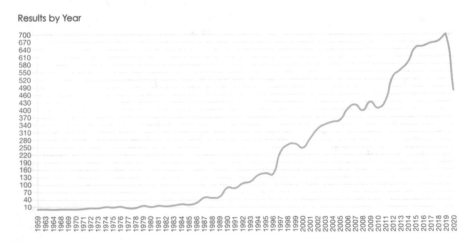

Fig. 9.2 Publication evolution of relevant interventional studies assessing an NPI in breast cancer prevention or care using Kalya Research (July 25, 2020). (Reprinted with permission from Kalya)

Fig. 9.3 Geographic distribution of relevant interventional studies assessing an NPI in breast cancer prevention or care using Kalya Research (July 25, 2020). (Reprinted with permission from Kalya)

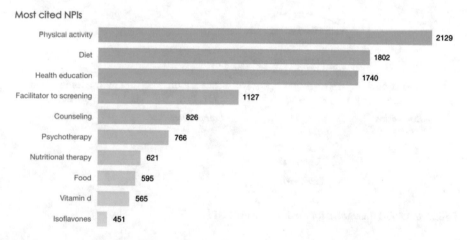

Fig. 9.4 NPIs assessed by relevant interventional studies assessing an NPI in breast cancer prevention or care using Kalya Research (July 25, 2020). (Reprinted with permission from Kalya)

ies reveal the natural use and receptive response of NPIs through methods such as retrospective analyses, prospective follow-ups, cohort studies, case-control protocols, case studies, focus groups, surveys, and qualitative interviews. Mechanistic studies reveal causal models through micro-analysis, bioactive compounds analysis (e.g., Yadav et al. 2020), in vitro experiments, and modelling methods. Interventional studies compare the benefits and risks of different outcomes using methods such as before-after studies, pilot studies, feasibility studies, controlled trials, randomized controlled trials, systematic reviews, and meta-analyses.

Most targeted outcomes

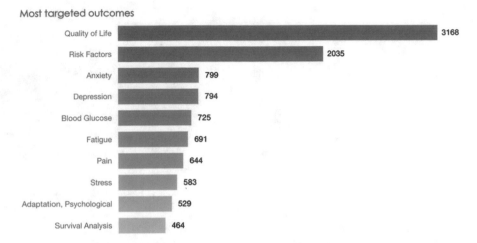

Fig. 9.5 Outcome measured by relevant interventional studies assessing an NPI in breast cancer prevention or care using Kalya Research (July 25, 2020). (Reprinted with permission from Kalya)

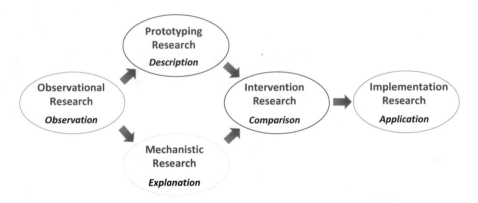

Fig. 9.6 Global framework for the assessment of NPIs

However, there are two categories of studies for NPIs that are neglected but important – prototyping studies and implementation studies (Fig. 9.6). Prototyping studies with specific methods, such as design prototyping methods, intervention characterizations, satisfaction surveys, user experience studies, ergonomics studies, and reliability tests are essential. Too many NPIs, such as connected health tools (e.g., smartphone applications for therapeutic education, serious games), are forgotten by patients after a few days or weeks of use. This category of research can legitimize NPIs by proactively co-constructing their best practices in conjunction with users rather than appearing ex nihilo and disconnected with future markets. This kind of NPI development emphasizes usability, privacy, and security protections that necessitate the systematic engagement of end users. Some authors proposed guidance on intervention development by providing step-by-step instructions for co-producing and prototyping a public health intervention's content and delivery processes (Hawkins et al. 2017).

Once the risks and effectiveness of NPIs are properly known, it is important to present the proper conditions for their implementation. While the general framework for NPI application must not change, conditions for adjusting to their context and delivery modes should be established and improved regularly. The resulting procedures, or best practices, thus become an integral part of the successful deployment of and practitioner training for NPI implementation. It is important to note that NPI application varies slightly between territories and cultures. As such, it is important to develop methods for the dissemination, surveillance, and analysis of NPIs with whistleblower investigations, impact follow-ups, big data analysis, intervention controls, and management practices.

4 Monitoring

4.1 Vigilance for NPI Use

From a medical point of view, biomedical products are closely monitored through various official control systems. General practitioners and specialists must report any drug problems and/or iatrogenic consequences to their respective national agencies. The French public health code recently invited patients to participate in monitoring, surveillance, and security actions for health interventions (article L4001-1, last modification of April 16, 2018). Yet, these feedbacks are still not required for NPIs, neither in France nor in Europe. NPIs are not considered as health products but rather remain as health solutions with risks. A vigilance system for NPI use that involves traceability and alert feedback is needed to detect risks and to guarantee the constant improvement of each intervention.

4.2 Professionals

Medical and paramedical practitioners have organizations that represent them: unions, colleges, orders, federations, societies, and associations. However, this is not the case for NPI professionals who are mostly freelancers, liberal professionals, or even volunteers. The word of mouth is their best asset, and the Internet is opening spaces for their expansion beyond borders, limits, and surveillance. Professions that are supposed to be beneficial to health should be identified, mapped, and potentially monitored (Dixon 2008).

A system should be created in order to assess these professionals on the basis of something more than their reputation or the number of likes on their social networks. These professionals should be required to complete initial and continuing training about implementations and ethics that are approved by authorities. It is necessary to identify abuses as soon as possible rather than waiting for the press to bring to light serious, irreversible, or dramatic cases.

> **Ethics**
> The obligation for honesty is written in both the ethical codes of medical professionals and in the French Public Health Act: "*they forbid charlatanism and deception, impose the prescription and distribution of treatments for which the efficacy was established. They also proscribe the use of obscure remedies or treatments which do not clearly list the substances that they contain*" (article 39 of the ethical code and article R.4127-39 of the Public Health law).

4.3   Practices

National, European, and international statistics on the practice of NPIs remain imprecise due to the lack of common denominators; on one hand, they escape the surveillance systems of healthcare administrations and on the other hand, patients do not always reveal their experiences to their doctor for different reasons (Mazzocut et al. 2016).

Economic, health, and scientific data are too fragmentary and approximate to generate understanding about the reality of professional practices and uses. It is time to put them into practice in individual life, care pathways, and data management systems to provide a more realistic appreciation of NPIs on the country, continental, and even global scales.

5   Regulation

5.1   Regulatory Status

NPIs are becoming increasingly included within healthcare professional routines. In some cases, they are even reimbursed for patients. Nevertheless, their regulation in Europe remains varied. For example, France does not yet have specific regulation for this sector (Wiesener et al. 2012), which seems to be nowhere and with no correspondence to consumer products and services. It does not correspond to invasive and biomedical treatments, either; instead, this sector remains in an intermediate position without clear rights for users and duties for practitioners and manufacturers (Chap. 1).

For the most part, NPIs are tolerated in many states, but they are not necessarily well supervised or even legal in some cases (e.g., the illegal practice of medicine or physiotherapy). There is a notable lack of sanctions that is justified in the name of free consumption as well as a lack of means for patients to distinguish real NPIs from fraudulent practices and competent professionals from charlatans that do not help. There is a known risk that recourse to some unregulated practices can act as a gateway to a sect or a fallacious company. There also remains a risk for NPIs to become confused with alternative remedies (Chap. 1).

The only way is to get NPIs out of the "not-not" category in which they are neither everyday products nor drugs, or out of the land of confusion (e.g., alternative medicines), to regulate the NPI ecosystem by giving them a real status.

The WHO (2013, 2019) has advocated for the integration of NPIs into conventional health systems with specific regulation. Its 2014–2023 plan invites member states to "(1) *strengthen the knowledge base for active management of traditional medicine / complementary medicine via appropriate national policies, (2) strengthen insurance quality, safety, appropriate use and effectiveness of traditional medicine / complementary medicine by regulating products, practices and practitioners, (3) promote universal health coverage by adequately integrating traditional medicine services / complementary medicine in the delivery of health services and self-care*" (WHO 2013, p. 44). Many politicians in Europe have been calling for better regulation since 1997, even if they have not made specific proposals for doing so. It is becoming urgent to clarify the status of NPIs as well as the conditions for their practice by professionals. Despite European directives, national agency reports, assembly or senate notes, government speeches, and policy-maker comments declared in the media, nothing significant has changed.

The Path of Politics?
On March 16, 1997, the European Union's Commission for the Environment, Public Health and Consumer Protection published a Report on the status of unconventional medicines. This Report urged the EU to "*match supply and demand on the basis of a double principle of freedom: freedom for patients to choose the therapy of their choice and freedom for practitioners to exercise their profession*" (European Parliament 1997, p. 11). A vote was made on a resolution on May 29, 1997 asking the European Commission to, if the results of the related examinations allowed, embark upon a process of recognition for non-conventional medicines and to take the necessary measures to promote the establishment of appropriate committees. In 2012, the first European research program on NPIs, called CAMbrella, identified three obstacles to this recognition: (1) The Treaties of Rome and Lisbon leave it to each state to organize its health system and its legislation; (2) there is a reluctance on the part of states to freely open up the healthcare offer and to harmonize it on European territory; (3) NPI organizations have already made progress on professional European harmonization and express reluctance toward other disciplines. So far, only 17 of the 39 countries in Europe have passed texts regulating complementary medicine.

Standardization represents a potential alternative to the political route toward NPI regulation. This would involve the provision of reference documents that are created collaboratively and consensually by all interested parties (Fig. 9.7). These documents would provide rules, characteristics, recommendations, and examples of good practice that are related to NPI products, services, methods, processes, and organizations. The NPIs ecosystem would benefit from standards that are designed under the responsibility of dedicated agencies such as the French Association for Standardization (AFNOR), the European Committee for Standardization (ECS), or the International Organization for Standardization (ISO). COVID-19 has shown how important it is to mobilize all actors in the face of an emergency or pandemic as a means of creating a benchmark for the manufacture of an effective mask against the virus. This "barrier mask" was made for the general public rather than as a medical device (Regulation EU / 2017/745) or as a form of personal protective equipment (Regulation EU / 2016/425). Its repository was developed on the initiative of the French standardization agency (AFNOR) in line with its mission and general interest. The resulting specifications were produced in the space of a week during the national confinement period and, thus, without any physical meeting. They were developed shortly after the WHO qualified COVID-19 as a pandemic. As such, this repository constitutes an in-depth example of a solution that was developed collaboratively in an emergency situation. The work to develop it was done in an open manner and was shared by a large number of actors, involving the collaboration of more than 150 experts. This type of procedure could be replicated for each NPI.

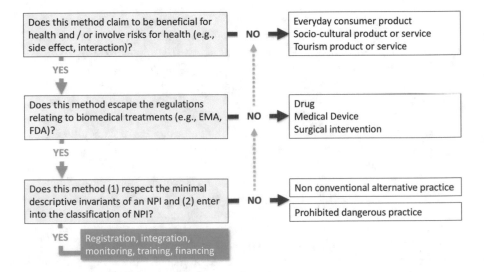

Fig. 9.7 A possible integration process of NPIs in health systems

A Path Toward Standardization?
The European Committee for Standardization (ECS) is a public standards organization that was founded in 1961 with a mission to foster the economy of the European Union through global trading and to guarantee the welfare of the 460 million European citizens through the development, maintenance, and distribution of coherent sets of standards and specifications. The 34 national members work together to develop European standards for various sectors such as food, safety, care, and health. They also work to build an internal European market for goods and services that helps to position Europe in the global economy. ECS is responsible for defining voluntary standards for different kinds of products, materials, services, and processes. It provides European standards and other technical documents that work to promote free trade, worker and consumer safety, network interoperability, environmental protection, research development and exploitation programs, and public procurement. Many ECS standards are coherent with International Organization for Standardization (ISO) standards.

5.2   Practitioners

According to the Public Health Code (consolidated version of April 16, 2018), the French State officially recognizes 24 "health professions" that do not include professionals such as psychologists, disease education teachers, exercise professionals, or osteopaths who are considered as lacking evidence of a direct effect on health. Yet, these health professionals claim to be distinguished from impostors, gurus…

It is time for the authorities to take this matter seriously and to provide clear directives. There remains too many legal loopholes and regulatory vagueness in this sector. We can no longer operate with a restrictive list of health professions that are totally unsuited to current health challenges and a pharaonic list of unregulated trades with no requirements for their training, ethics, knowledge, implementation, aptitude, rights, or responsibilities in practice.

Internet Ubiquity
Following a series of hearings, the French Senate noted that alternative medicine practitioners *"canvass the public in the classified ads of certain free newspapers, in wellness fairs and other congresses of quantum medicine. They are especially ubiquitous on the Internet"* (French Senate 2013, p. 35).

Regulations should be put into place for NPI trades so that they come with ethics codes, contractual charters, and strong commitments for their contribution to research and the advancement of evidence-based practice. In a rapidly changing world involving dramatic environmental, demographic, and epidemiological changes, there is room for everyone.

Drawing inspiration from the pharmaceutical field, a transversal organization that federates NPI professionals should be created. An NPI alliance should be created as a non-profit trade association of stakeholders that are engaged in the evidence-driven advancement of NPI growth, development, and use. This alliance could work to expand access to high-quality, evidence-based NPIs for patients, healthcare providers, and financers in order to improve their clinical, health, and economic outcomes. Its mission would be to understand, adopt, and integrate clinically validated NPIs into healthcare prevention and care through education, advocacy, and research. In this vein, the Society for Integrative Oncology produced an evidence-based guideline on use of NPIs during and after breast cancer treatment that was determined to be relevant to the American Society of Clinical Oncology (Lyman et al. 2018).

> **A European Drug Federation Based on Common Denominators**
> The European Federation of Pharmaceutical Industries and Associations (EFPIA) represents the biopharmaceutical industry operating in Europe. Its mission is to create a collaborative environment that enables its members to innovate, discover, develop, and deliver new therapies and vaccines for people across Europe, as well as to contribute to the European economy. The EFPIA created a manifesto sharing its vision along with proposals on how to build a healthier future and brighter tomorrow for Europe. This document allows the research-based pharmaceutical sector to advocate for a healthier life as a priority for its citizens, for a better patient-centered health system, and for a pro-innovation environment.

5.3   Prescription

If nothing will stop the self-prescription of NPIs, shouldn't doctors be more interested in NPIs so that they can prescribe them if necessary? Doctors must be convinced of what NPIs are, realize that they use them themselves, become aware of what they are, and be trained in their proper use.

Indeed, general practitioners (GP) are becoming increasingly interested in the prescription of NPIs. More than half of office-based physicians recommended at least one NPI to their patients in the United States (Stussman et al. 2020). These doctors see their patients more regularly than specialists and they are more attentive to prevention. The fact that many GPs are gathering in multi-professional health centers in proximity with other care and prevention specialists will accelerate the movement.

An essential part of the doctor's role is to thoroughly question the patient's intention for NPI use, allowing them to be informed about what the patient hopes to address in their NPI use and to help orient them toward the best available solutions. In the absence of data on efficacy or risk, doctors should not reject patient requests, which thus risks breaking their relationship of trust. Instead doctors can use search tools, such as Internet databases like PubMed, Science Direct, Central, or Motrial, to look for potential drug interactions with the patient's biomedical treatment. Only if there are no assumed risks can doctors authorize NPI use, which they should monitor as long as the patient is not involved in a clinical trial.

Focus

The Collège des Médecins du Québec has established a code of ethics containing rules to be adopted when faced with unrecognized treatments, including complementary and alternative medicine. According to this code, "the doctor must, with respect to a patient who wants to use insufficiently proven treatments, inform him of the lack of scientific evidence relating to such treatments, of the risks or disadvantages that could result from it, as well only benefits that would provide him with usual care, if any." The doctor, therefore, has the responsibility to inform their patient on the basis of evidence gathered according to a proven scientific protocol. When a patient decides to opt for an NPI practice, the doctor must ensure follow-up. The College also insists on an exchange of information from the patient to the doctor so that the latter is aware of the alternative methods being used. Likewise, each practitioner is held legally responsible for the care they provide. Thus, a physician who refers to or advises the use of an NPI practice can be found guilty of negligence or professional misconduct if their advice is not based on current scientific evidence. Doctors are not, however, held responsible for any faults committed by NPI therapists (Gaboury et al. 2016).

5.4 Reimbursement

New offers for NPI reimbursement are now being made by complementary health insurance companies, insurance companies, provident organizations, and even banks.

The WHO report (2013) on complementary medicines describes examples from China, Japan, and Switzerland in which NPIs are reimbursed by the national health insurance if they follow a medical prescription. In parallel, healthcare spending is increasingly weighing on state economies (Chap. 7) as biotechnologies become more and more expensive. For example, a single innovative chemotherapy can cost around US\$30,000. In this era of limited resources, state health insurance can no longer be the only financial source to meet patient demand.

The advantage of NPIs is that they cost far less to validate, monitor, and offer patients than biotechnology products.

Cost Savings with NPIs

Some NPIs reduce health insurance expenditures (fewer hospitalizations, less toxic treatments, less polytherapy, less sick leave) and patient expenditures (fewer dependents). On average, 1 euro invested in NPIs saves 3 when people are at risk of disease and 5 when people are chronically ill. In addition, this investment creates jobs in innovation, personal services, and remote monitoring. The only drawback is that funding is currently short-term rather than long-term, reimbursements remain partial, and many procedures remain complex or unknown.

Medico-economic studies in the NPI sector indicate a rationale for reimbursing psychotherapies, osteopathic methods, and smoking cessation methods. By including NPIs within reimbursable health pathways, in particular, for people suffering from noncommunicable diseases, insurance can demand results from NPI professionals and patients.

By implementing NPIs in this way, healthcare financers will encourage us to think in terms of positive health outcomes and multi-generational incentives for prevention. They can reward good health behaviors with bonuses, rather than punish and stigmatize bad ones. They can facilitate the development of patient empowerment by helping them to take care of themselves. As such, it will no longer be a question of demonizing this or that health solution – such as what we read in certain books written by doctors ("happiness without drug prescription," for example) – but rather of taking the best of both worlds and covering unmet medical needs.

Prevention Transformation

Prevention can involve something other than deprivation if it helps to heal or prevent illness. We should stop making it a constraint or an attack on individual freedom; instead, we should make it a springboard for empowerment. The WHO Montevideo 2018–2030 roadmap on noncommunicable diseases makes this a priority for sustainable development: *"We [Heads of State and Government, Ministers and Representatives of the States and Governments participating in the Montevideo conference from 18 to 20 October 2017] will give priority to the best interventions which are satisfactory, affordable, fair and cost-effective, based on factual data and adapted to national priorities and contexts, which will allow obtaining the highest return on investment. We will focus on health as a political priority, through measures to combat the impact of the main risk factors of noncommunicable diseases, such as regulation, standard setting and fiscal policies, among others, respecting the internal legal frameworks of the countries and their international obligations"* (WHO 2017, Resolution 7, p. 3).

This also means that we should avoid an unequal treatment for NPIs. These solutions do not question the merits of medicines or medical devices. Instead, they help to alleviate the negative impact of biomedical treatments and to improve patient health and quality of life. NPIs utilize other levers to maximize the chances of patient recovery and to reduce their suffering. Unfortunately, studies such as Kemppainen et al. (2018) indicate that there are social inequalities in European access to and use of NPIs.

5.5 A Unique Validation and Surveillance Framework

Currently, we are not witnessing a convergence toward a consensual framework for the validation and monitoring of NPIs. Rather, we are seeing the multiplication and diversification of models (Chap. 8). For example, a narrative review listed the existence of 46 frameworks as of April 2019 (Carbonnel and Ninot 2019). Debate rages among the partisans of a minimal surveillance who are emphasizing the low risks of NPIs and the freedom of choice, the partisans advocating for an NPI integration involving the relevance factors established by groups of experts, and the partisans of a strict audit and safety monitoring of NPI validation before they go to market (efficacy, risks, target population).

The adoption of a single model for validating and monitoring NPIs on a continent-wide or even a global scale would make NPI assessment procedures more consistent for researchers, more secure for users, more transparent for decision makers, and more programmable for entrepreneurs. The pharmaceutical industry did this 50 years ago (Van Norman 2016a); the medical devices sector did the same in 2018 (Van Norman 2016b); it is now high time for the NPI sector to follow suit.

5.6 Intellectual Property

A major problem that needs to be resolved concerns the issue of intellectual property and NPIs. Many of these practices rely on know-how that is often passed down orally from teacher to student, or on techniques and processes that are easy to copy and difficult to protect by regulation.

In addition, the abundance of free globalized digital information sharing via generalist or specialized health-related sites/apps are facilitating plagiarism without geographical and temporal limits. This reduces the dissemination of good practices and neglects the recognition of the original work of NPI creators and promoters. As a result, this difficulty reduces innovation and investment in additional research and development.

Actions should be taken by policy makers to help creators, researchers, developers, engineers, and practitioners to better protect their innovations in all territories and for sufficiently long periods. It is not enough to register a trademark or a software code; patents are often poorly applied. Stricter regulations would enhance the development, evaluation, and continuous improvement of NPIs.

> **Protect the Intellectual Property Rights**
>
> *As traditional and complementary medicine becomes more popular, it is important to balance the need to protect the intellectual property rights of indigenous peoples and local communities and their health care heritage while ensuring access to T&CM and fostering research, development and innovation. Any actions should follow the global strategy and plan of action on public health, innovation and intellectual property* (WHO 2013, p.19).

6   Training

NPIs are methods that should be applied by willing, informed, and responsible individuals who are supervised by a qualified professional. Professionals must learn to collaborate with NPI users, but these kinds of technical alliance skills and know-how cannot be decreed. Rather, they must be learned. Doing so fosters the conditions for generating better health benefits, for improving patient adherence, and for reducing adverse effects (Chap. 5).

The use of NPIs requires a human contribution throughout the explanation, decision-making, monitoring, and treatment assessment process. Professionals must give patients the means to listen to their physical and psychological signs, to be attentive to auto-aggression, to reflect on their disagreements and misunderstandings, to express their feelings, and to ask for help in case of difficulty. To do so, professionals must create "*a warm emotional bond or collaborative attachment with the patient*" (Flückiger et al. 2018, p. 333).

> **Favorable Conditions for a Successful Therapeutic Alliance**
>
> - Material (e.g., location, furniture, equipment, hygiene, odor, lighting, atmosphere)
> - Psychological (e.g., attention, availability, empathy, kindness, authenticity)
> - Educational (e.g., explanation without technical jargon, relevant analogy, metaphor, good interpersonal distance, didactic progressiveness)
> - Ethics (e.g., respect for dignity, confidentiality of words and data, proportionate alert, networking, traceability of acts)

These skills are refined by both initial and continuous training because experience is not enough to become an expert. Fostering an effective therapeutic alliance requires technical skills in order to:

- Eliminate concerns about a health problem and a preventive and/or therapeutic solution
- Guarantee maximum safety conditions (e.g., emergency procedures, first aid, monitoring of health alerts, vaccination, hygiene)
- Facilitate the patient's active participation by proposing open-ended questions, seeking their point of view, allowing sufficient time for responses, and discussion
- Encourage the patient's active participation in the discussion and management of their own health
- Integrate NPI use for the specific health purpose for which it was created and validated
- Help patients develop their own strategies to achieve their desired goals
- Collaborate with patient's family
- Work in a multidisciplinary team
- Contribute to advancing practices through research

In the same vein, NPI professionals must know how to stay within the bounds of their defined role rather than playing medical doctor if they are not qualified as one. Upon the initiation of an NPI practice, the hypothesis of a serious illness should never be forgotten and should be considered at the slightest sign or doubt. Additional medical examinations must be requested as necessary. In terms of NPI professionals, this justifies the need for a minimum knowledge of physiopathology and human psychopathology within their training.

Patient Expectations Beyond Evidence-Based Interventions with High Levels of Control for Quality, Delivery, and Reliability

- An initial meeting long enough to discuss the details of the patient's history
- Attentive listening
- A favorable relational climate created in a soothing place
- A dialogue on how to improve patient lifestyle
- Help to reduce daily stress
- An explanation of how the manualized NPI will mobilize the patient's internal resources for health improvement
- Assurances that the NPI will help to resolve the patient's health over time
- A benevolent contact
- Regular monitoring
- Security
- Long-term privacy and secure data protection
- Implementing NPI(s) use in an engaging and convenient way

6.1 Initial Training

The initial training of NPI practitioners can no longer remain isolated in different schools of thought. Far too many of the training programs outside of the medical professions do not provide enough knowledge on pathophysiology, psychopathology, evidence-based practice, health and safety procedures, the healthcare system, and multidisciplinary teamwork.

Training represents one of the keys to improving their institutional recognition and credibility. Ministries of Health, Education, and Research should work together to ensure the quality and harmonization of NPI training diplomas and course content. This would allow these practices to be integrated within undergraduate, postgraduate, and continuing professional health and prevention curricula. Along these lines, an academic college of integrative and complementary medicine (CUMIC) was created in France in 2018 as a means of accelerating the development of quality and homogeneity in French NPI training courses in medical and non-medical trades.

Progress in the Initial Training of Trades Related to NPIs Still Needed
General academic education curricula for health and health care remain too theoretical and too abstract. Private professional curricula remain too applied and disconnected from evidence-based practice. Some schools openly oppose national or international recommendations. Training supervisors are rarely compensated for the entirety of their work (trainee follow-up, educational meetings, report feedback, oral exams). Debates remain over the levels and years necessary for diplomas. Universities argue for the 3-year license and the 5-year master degree. Private schools prefer 2- and 4-year curricula. Virulent debates also surround the consideration of NPIs as an optional or mandatory course within the medical curriculum. In short, an immense amount of work remains to be done in order to harmonize the initial trainings for NPI professionals, both in terms of teaching content and the recognition of diplomas by employers and authorities.

Another dark spot concerning NPI initial trainings concerns research experience – especially clinical and translational research. Most prevention and healthcare professions do not receive proper overviews of the validation and implementation process for NPIs within their curriculum, which is either too ad hoc or too targeted. For example, an American nurse can obtain a PhD in nursing, but this diploma does not exist for nursing in France. French nurses are thus confined to an assistant role and are given limited opportunity to advance their careers. As such, including research and assessment on NPIs in healthcare curriculum represents an essential means of bringing to light what should be maintained and what should be personalized in clinical practice. Promoting closer collaboration of students and researchers represents an excellent means of helping dangerous practices to disappear and to advance best practices in the NPI sector. Evidence-based practice necessitates that NPI professionals complete initial trainings that allow them to contribute to scientific research protocols. Doing so constitutes an excellent way to establish the rigor necessary for their evaluation and professional practice. Furthermore, collaboration with academic research departments should also be encouraged, as was suggested by the WHO (2013) and the European Commission (1997).

> **The European Parliament Initiative of 1997**
> *To support and legitimize the process of recognition of non-conventional medical disciplines, we must go further: organize dialogue between the academic community and the experts in each discipline; conduct multidisciplinary research programs based on jointly defined methodologies and appropriate validity criteria* (European Parliament 1997, p. 13).

6.2 Continuing Training

In the field of professional aviation, pilots are not allowed to fly an aircraft until they have obtained the specific qualification and license ensuring suitability controls. When renting a machine intended for construction, a short training session is required in order to ensure its proper use and to minimize the risk of accidents. Professionals claiming to have an effect on human health through the practice of NPIs, however, do not currently need specific training and are not required to complete a mandatory training session. This begs the question of whether health is merely a consumer good like any other. Shouldn't users be privy to knowledge about their practitioners' qualifications and certifications? Considering the exponential acceleration of NPI research and innovation worldwide, there is a need for maintaining technical expertise through certification and continuing education programs that are regularly controlled by peer experts. In the field of aviation, this reliable, evidenced-based approach ensuring a high level of quality control has proven its worth for user safety. The same should be true for NPIs.

Going beyond continuing education for NPI training, there is a need for ensuring the continuous provision of information on NPIs for doctors, surgeons, pharmacists, dental surgeons, and administrative managers. As doctors receive more and more requests from their patients for NPIs that they have read about on the Internet, they must stop evading the question or condemning the practice, which only pushes their patients to abandon medical treatment under the pretext of conspiracy theories about pharmaceutical industrialists who are radically opposed to natural medicines.

Information for Healthcare Professionals
Improve the adherence of health professionals to the recommendations on non-pharmacological therapies (French National Authority for Health 2011, p. 52).

6.3 Digital Dissemination

It is certain that the future healthcare system will look very different from what we know today. Trends such as increasing costs, value-based payment models, and personalized care are all coming together to disrupt traditional medical activities. Healthcare professionals must be better aligned with the future of health care. For example, this could occur through videos that transmit experiences that are difficult to explain in words. These could share the inexpressible and the inexplicable through non-verbal communication. They would also allow patients, professionals, engineers, and researchers to better understand what NPIs are in both their technicality and their complexity. Digital technologies encourage caregivers and preventionists to foster targeted NPI training, to learn efficient information-sharing techniques, and to participate in collaborative team meetings without geographical limits.

An Example of Healthcare Information Technology
The English have made progress on this path by creating "HealthTalk (2020)," where patient narratives are verified by experts before being made public.

6.4 Actients

Some people have been living with a chronic disease for 50 years or more (e.g., asthma, type 1 diabetes, mental disorder). Over time they learn to manage their disease, and many aspire to share their experiences with peers and professionals. For the past 10 years, initiatives in therapeutic education and in chronic patient networks (e.g., AIR + R Network in French Occitanie Region) have made it possible for knowledge on good strategies for coping with chronic disease to accumulate and be shared; many different NPIs are included in these information repositories. These forms of profane, experiential knowledge about NPIs and their delivery models deserve to be transmitted. In so doing they transform the meaning of the word "patient" into the word "actient."

The "Actient" Instead of the "Patient"
This neologism gains merit through its emphasis of the notion of an "actor." Many patients want to become active and to learn from their peers through dedicated sessions. Some become experts in their illness and are capable of speaking about their conditions better than anyone else through their sharing of practical solutions that prioritize their needs. Creating places to learn and train, like the Sorbonne Patient University, is also a remarkable innovation in this domain. The notion of "patient partner" has been born. Following the initiative of Professor Christian Préfaut, the Montpellier Faculty of Medicine was one of the first French medical universities to offer young doctors courses given by chronically ill patients. This allowed students to meet people with things to share rather than being restricted to interactions with patients in acute crisis in a hospital where speech is restricted and conditioned.

6.5 Family Caregivers

Alzheimer's disease provokes a progressive burden for patient families that goes beyond financial considerations to influence their physical, relational, and emotional lives. About 20% of the French population cares for a dependent person or a person with a loss of autonomy in order to keep them at home. About 10% provide daily practical, moral, and financial support to people over 60 wishing to live at home. These family caregivers are important vectors of NPI practices both for the people they help and for themselves.

Who Are Family Caregivers?
A family caregiver is a non-professional who supports, on an ad hoc or permanent basis, a dependent person in their activities of daily living. This help may include care, care coordination, health education, social assistance, administrative procedures, alertness, psychological support, and/or household chores. In many cases, this responsibility weakens the caregiver both in their personal and professional lives. They become burdened by fears of not doing enough and the guilt of doing something wrong. Family caregivers in France are 60 years old on average; 47% are employees and 60% are women. Their burden is such that 48% of them end up getting sick in turn. The law number 0301 (December 29, 2015) recognizes and regulates the status of family caregivers. In particular, it gives caregivers the right to respite, facilitating access to adapted stays. It also provides support measures for salaried caregivers so that they can take leave without a loss of wages, such as is the case in Italy and the Netherlands.

One in two family caregivers express an interest in care training, but only 1 in 10 actually receive it. Training them is a necessary means of helping them to better understand the contours of their mission and the needs of fragile people; to choose the right solutions – especially NPIs – to better understand the recommendations for good practice; to assess their limits; to seek help before exceeding their capacity; and to know their rights. Digital NPI practices (domestic solutions, remote assistance, home automation, applications) and ergonomic solutions (armchairs, beds, etc.) are being developed to prevent and assist in cases of frailty.

7   Conclusion

The future of NPIs depends on improving transparency and content of NPI interventions so that they consistently meet applicable evidence-based requirements and specifications. This would require practitioners to complete specific qualification sessions. It would also necessitate stronger connections among scientific progress, practitioners, and prescribers in order to ensure the meaningful delivery of NPIs. To do so, improvements in the standardization and provision of initial and continuing education training are needed. Furthermore, the dissemination of NPIs would be greatly improved and facilitated through the creation of a quality labeling system and a real regulatory status that protects users (especially children and vulnerable people) and professionals. As such, the creation of a transnational organization would be a real advantage to better shape this emerging ecosystem, especially to differentiate non-pharmacological interventions from unconventional alternative interventions (e.g., alternative medicine), sociocultural interventions (e.g., art), and dangerous interventions (e.g., sectarian activity, dangerous products).

Key Points
The future of NPIs depends on:

- Transparent and standardized descriptions in a best-practice repository
- NPI-related qualifications for practitioners
- A greater awareness among biomedical practitioners and financers
- Improved initial and continuing training that is based on high quality and evidence-based practice
- The creation of regulations facilitating NPI integration into individual patient health and life pathways
- The adoption of a common validation and monitoring framework
- The increased participation of users and professionals in NPIs assessments
- The creation of a quality label to improve NPI recognition within the healthcare system
- An international regulatory status that protects users and professionals
- More reimbursements
- The creation of an organization that goes beyond siloed operations in order to produce expert opinions, recommendations, and alerts in the emerging NPI ecosystem

References

Australian Center in Complementary and Integrative Medicine. (ARCCIM): https://www.uts.edu.au/research-and-teaching/our-research/complementary-and-integrative-medicine.

Beck, K., Beedle, M., Bennekum, A., Cockburn, A., Cunningham, W., Fowler, M., Grenning, J., Highsmith, J., Hunt, A., Jeffries, R., Kern, J., Marick, B., Martin, R., Mellor, S., Schwaber, K., Sutherland, J., & Thomas, D. (2001). Manifesto for agile software development. www.agilemanifesto.org.

Boutron, I., Moher, D., Altman, D. G., Schulz, K. F., Ravaud, P., & Group, C. (2008). Methods and processes of the CONSORT group: Example of an extension for trials assessing nonpharmacologic treatments. *Annals of Internal Medicine, 148*(4), W60–W66.

Boutron, I., Ravaud, P., & Moher, D. (2012). *Randomized clinical trials of non pharmacological treatments*. Bacon Raton: CRC Press Taylor and Francis.

Carbonnel, F., & Ninot, G. (2019). Identifying frameworks for validation and monitoring of consensual Behavioral intervention technologies: Narrative review. *Journal of Medical Internet Research, 21*(10), e13606. https://doi.org/10.2196/13606.

Clarke, A. E., Mamo, L., Fosket, J. R., & Fishman, J. R. (2010). *Biomedicalization: Technoscience, health, and illness in the US*. Edinburgh: Duke University Press.

Cochrane collaboration for Complementary Medicine (2020). https://cam.cochrane.org/

Cohen-Mansfield, J. (2018). Non-pharmacological interventions for persons with dementia: What are they and how should they be studied? *International Psychogeriatrics, 30*(3), 281–283. https://doi.org/10.1017/S104161021800039X.

Collins, L. M., Murphy, S. A., & Strecher, V. (2007). The multiphase optimization strategy (MOST) and the sequential multiple assignment randomized trial (SMART): New methods for more potent eHealth interventions. *American Journal of Preventive Medicine, 32*(5), S112–S118. https://doi.org/10.1016/j.amepre.2007.01.022.

Confederation of Practitioners in Natural Medicine of Quebec (2020). https://cpmdq.com/

Consortium on Integrative Medicine (2020). https://imconsortium.org/.

Craig, P., Dieppe, P., MacIntyre, S., Michie, S., Nazareth, I., & Petticrew, M. (2008). Developing and evaluating complex interventions: The new Medical Research Council guidance. *British Medical Journal, 337*, a1655. https://doi.org/10.1136/bmj.a1655.

Czajkowski, S. M., Powell, L. H., Adler, N., Naar-King, S., Reynolds, K. D., Hunter, C. M., et al. (2015). From ideas to efficacy: The ORBIT model for developing behavioral treatments for chronic diseases. *Health Psychology, 34*(10), 971–982. https://doi.org/10.1037/hea0000161.

Danell, J. B., Danell, R., & Vuolanto, P. (2019). Scandinavian research on complementary and alternative medicine: A bibliometric study. *Scandinavian Journal of Public Health*, 1403494819834099. https://doi.org/10.1177/1403494819834099.

De Grey, A., & Rae, M. (2007). *Ending aging. The rejuvenation breakthroughs that could reverse human aging in our life time*. New York: St Martin's Press.

Dixon, A. (2008). *Regulating complementary medical practitioners*. London: King's Fund.

European Parliament. (1997). *Rapport sur le statut des médecines non conventionnelles. Commission de l'environnement, de la santé publique et de la protection des consommateurs*. Brussels: European Parliament.

Flückiger, C., Del Re, A. C., Wampold, B. E., & Horvath, A. O. (2018). The alliance in adult psychotherapy: A meta-analytic synthesis. *Psychotherapy, 55*(4), 316–340. https://doi.org/10.1037/pst0000172.

French National Authority for Health. (2011). *Développement de la prescription de thérapeutiques non médicamenteuses validées*. Paris: HAS.

French Ministry of Health and Solidarity. (2018). *2018–2022 French health strategy*. Paris: French Ministry of Health and Solidarity.

French Senate. (2013). *Rapport 480 au nom de la commission d'enquête sur l'influence des mouvements à caractère sectaire dans le domaine de la santé*. Paris: Senate.

Gaboury, I., Johnson, N., Robin, C., et al. (2016). Médecines alternatives et complémentaires: Les médecins se considèrent-ils en mesure de répondre aux exigences du Collège des médecins du Québec? *Canadian Family Physician, 62*(12), e767–e771.

Harari, Y. N. (2017). *Homo Deus. A brief history of tomorrow.* Toronto: McClelland & Stewart.

Hawkins, J., Madden, K., Fletcher, A., et al. (2017). Development of a framework for the co-production and prototyping of public health interventions. *BMC Public Health, 17*(1), 689. https://doi.org/10.1186/s12889-017-4695-8.

Hayles, K. (1999). *How we became posthuman, virtual bodies in cybernetics, literature and informatics.* London: The University of Chicago Press, Ltd.

HealthTalk (2020). www.healthtalk.org

Hoffmann, T.C., Erueti, C., Glasziou, P.P. (2013). Poor description of non-pharmacological interventions: analysis of consecutive sample of randomised trials, 347, f3755. https://doi.org/10.1136/bmj.f3755

Hoffmann, T. C., Walker, M. F., Langhorne, P., Eames, S., Thomas, E., & Glasziou, P. (2015). What's in a name? The challenge of describing interventions in systematic reviews: Analysis of a random sample of reviews of non-pharmacological stroke interventions. *British Medical Journal Open, 5*(11), e009051. https://doi.org/10.1136/bmjopen-2015-009051.

ICEPS Conference (2020). www.icepsconference.fr.

Irish Cancer Society. (2018). *Understanding cancer and complementary therapies.* Dublin: Irish Cancer Society.

Kalya Research (2020). www.kalya.ai.

Kemppainen, L. M., Kemppainen, T. T., Reipaninen, J. A., Salmenniemi, S. T., & Vuolanto, P. H. (2018). Use of complementary and alternative medicine in Europe: Health-related and sociodemographic determinants. *Scandinavian Journal of Public Health, 46*, 448–455. https://doi.org/10.1177/1403494817733869.

Kurzweil, R. (2005). *The singularity is near: When humans transcend biology.* New York: Penguin.

Lyman, G. H., Greenlee, H., Bohlke, K., Bao, T., DeMichele, A. M., Deng, G. E., Fouladbakhsh, J. M., et al. (2018). Integrative therapies during and after breast cancer treatment: ASCO endorsement of the SIO clinical practice guideline. *Journal of Clinical Oncology, 36*(25), 2647–2655. https://doi.org/10.1200/JCO.2018.79.2721.

Maryland Integrative Medicine Center (2020). http://cim.umaryland.edu

Mazzocut, M., Truccolo, I., Antonini, M., Rinaldi, F., Omero, P., Ferrarin, E., De Paoli, P., & Tasso, C. (2016). Web conversations about complementary and alternative medicines and cancer: Content and sentiment analysis. *Journal of Medical Internet Research, 18*(6), e120. https://doi.org/10.2196/jmir.5521.

Medoucine (2020). https://www.medoucine.com/

Motrial (2020). www.motrial.fr

National Center for Complementary and Integrative Health (NCCIH) (2020). www.nccih.nih.gov.

Niri (2020). www.niri.fr.

Nishare (2020). www.nishare.fr.

Norway's National Research Center in Complementary and Alternative Medicine (NAFKAM) (2020). http://nafkam.no/en.

Oude Engberink, A., Badin, M., Serayet, P., Pavageau, S., Lucas, F., Bourrel, G., Norton, J., Ninot, G., & Senesse, P. (2017). Patient-centeredness to anticipate and organize an end-of-life project for patients receiving at-home palliative care: A phenomenological study. *BMC Family Practice, 18*(1), 27. https://doi.org/10.1186/s12875-017-0602-8.

Powell, L. H., Freedland, K. E., & Kaufmann, P. G. (2020). *Behavioral clinical trials for chronic diseases: Scientific foundations.* New York: Springer.

Schwartz, D., & Lellouch, J. (2009). Explanatory and pragmatic attitudes in therapeutical trials. *Journal of Clinical Epidemiology, 62*(5), 499–505. https://doi.org/10.1016/j.jclinepi.2009.01.012.

Skelly, A. C., Chou, R., Dettori, J. R., Turner, J. A., Friedly, J. L., Rundell, S. D., Fu, R., Brodt, E. D., Wasson, N., Kantner, S., & Ferguson, A. J. R. (2020). *Noninvasive nonpharmacological treatment for chronic pain: A systematic review update (20-EHC009)*. Rockville: Agency for Healthcare Research and Quality. https://doi.org/10.23970/AHRQEPCCER227.

Stanford Center for Integrative Medicine (2020). https://stanfordhealthcare.org/medical-clinics/integrative-medicine-center.html

Stevens, A. (1881). *Madame de Staël: A study of her life and times, the first revolution and the first empire*. London: Harper & Brothers.

Straus, S., Glasziou, P., Richardson, W. S., & Hayne, S. B. (2019). *Evidence-based medicine: How to practice and teach EBM* (5th ed.). China: Elsevier.

Stussman, B. J., Nahin, R. R., Barnes, P. M., & Ward, B. W. (2020). U.S. physician recommendations to their patients about the use of complementary health approaches. *Journal of Alternative and Complementary Medicine, 26*(1), 25–33. https://doi.org/10.1089/acm.2019.0303.

University of Texas (2020). https://www.mdanderson.org/patients-family/diagnosis-treatment/care-centers-clinics/integrative-medicine-center.html

Van Norman, G. A. (2016a). Drugs, devices, and the FDA: Part 1: An overview of approval processes for drugs. *JACC: Basic to Translational Science, 1*(3), 170–179. https://doi.org/10.1016/j.jacbts.2016.03.002.

Van Norman, G. A. (2016b). Drugs, devices, and the FDA: Part 2: An overview of approval processes: FDA approval of medical devices. *JACC: Basic to Translational Science, 1*(4), 277–287. https://doi.org/10.1016/j.jacbts.2016.03.009.

WHO. (2013). *WHO traditional medicine strategy: 2014–2023*. Geneva: World Health Organization.

WHO. (2017). *Montevideo roadmap 2018–2030 on NCDs as a sustainable development priority*. Geneva: World Health Organization.

WHO. (2019). *Global report on traditional and complementary medicine*. Geneva: World Health Organization.

Wiesener, S., Kalkenberg, T., Hegyi, G., Hök, J., Roberti di Sarsina, P., & Fønnebø, V. (2012). Legal status and regulation of complementary and alternative medicine in Europe. *Forschende Komplementmedizin, 19*(S2), 29–36. https://doi.org/10.1159/000343125.

Yadav, S. K., Ir, R., Jeewon, R., et al. (2020). A mechanistic review on medicinal mushrooms-derived bioactive compounds: Potential mycotherapy candidates for alleviating neurological disorders. *Planta Medicine*, https://doi.org/10.1055/a-1177-4834

Chapter 10
Conclusion

Most complementary and alternative medicine (CAM) or traditional and complementary medicine (T&CM) products and practices are not assessed scientifically. This refusal is based on the pretext that they are natural, traditional, or mysteriously beneficial. However, progress in scientific methods, interventional studies, and data-collection systems are now available to explain and determine benefits and risks in targeted populations. Consumers choosing these remedies should be calling them *alternative medicines* and should be aware that they are using them at their own risk. Some are recreational, some are dangerous (Chap. 5); whatever the case, they definitely do not belong to evidence-based health interventions driven by high quality practice, delivery, and surveillance. Non-invasive human interventions with significant effectiveness on health, autonomy, quality of life, and longevity should be called *non-pharmacological interventions* (NPIs).

1 Science as Justice of the Peace

NPIs have chosen another path: longer, more tedious, but safer for users and for society – that of science and evidence-based practice (Chap. 8). A sufficiently detailed non-pharmacological method targeting a specific health objective for a given population can be evaluated and show a similarity of results in an equivalent context (Chap. 1).

This path was traced by the standard required for new drugs about 50 years ago. With methodological adjustments, including greater detail regarding their specificity, it is evident that more and more numerous and rigorous studies appearing since the beginning of this century prove the efficacy regarding health, safety, and economic usefulness of certain NPIs (Chap. 4). Comprehensive, systematic, and convergent compilations of these studies consolidate opinions. Sophisticated devices help understand their mechanisms of action: plural, simultaneous, chain, and systemic (Chap. 3).

Many NPIs prevent disease, change behaviors harmful to health, improve health, and increase lifespan in good health. Some, fewer in number, treat health problems that handicap the daily lives of patients. Although rare, some cure illnesses. By betting on evaluation by science, NPIs perfect themselves and become increasingly specialized. Their professionals are progressing. They do not aim to replace conventional biotechnological treatments, but to complement them, to humanize them, and in some cases to potentiate them.

NPIs are beginning to find their place in personalized care and healthcare routes, abandoning the derogatory qualifiers pronounced by their detractors, such as comfort care, supportive care, magic remedies, etc. (Chap. 2). Some countries reimburse them. A vast global market, growing by more than 5% each year, is opening up and will necessarily expand with the aging of the population and the increasing number of people with chronic disease (Chap. 7).

This book is a testament to advances in science and clinical research in the area of NPIs. The era of optimizing techniques, doses, content, training, and certified trades has come. This is the only path that will eliminate abusive practices, fraudulent products, impostors, and charlatans. It is also the only path, in the name of medical pluralism and freedom of choice that will make it possible to meet users' expectations (Chap. 6), to enable them to know their chances of getting better, the efforts they need to make, and the risks involved.

2   NPI Ecosystem

Science is nothing without law and political decisions. NPIs must get out of their nebulous legal position and have appropriate regulatory status between everyday consumer products and biotechnological therapeutics. From this position, decision makers and health authorities will be able to more easily regulate practices, user information, and professional training. A transnational organization is needed to structure this emerging ecosystem (Chap. 9).

Public and private insurance and provident systems have seized on the interest in NPIs to respond to the epidemiological, demographic, digital, and ecological transitions that lie ahead. The question now is how to finance them. Everything suggests that reimbursements will not be made without careful consideration. Users will need to commit to taking them, reporting side effects, and reporting abuse. Health is precious, so it's important to be careful to use solutions that effectively protect it.

3   Health Is Well-being: WHO's Utopia Becomes Reality

At a time when transhumanism wishes to make humans immortal and hybridized with robots, NPIs make the utopia proposed by the WHO in 1946, of human health defined as "a state of complete physical, mental and social well-being" and not merely an "absence of disease or infirmity," more salient.

NPIs show the way toward a quest for meaning for oneself, for others, and for the planet. They take us away from the path that makes us beings made up of clusters of replaceable cells and fixed hereditary predispositions. It's up to us to choose our positive health journey!

Appendix: Useful Websites

Research

ClinicalTrials.gov	clinicaltrials.gov
Clinical Trials Registry India (CTRI, India)	ctri.icmr.org.in
Cochrane Library	cochranelibrary.com
Cochrane Complementary Medicine	cam.cochrane.org
Consolidated Standards of Reporting Trials (CONSORT)	consort-statement.org
EU Clinical Trials Register (ECTR, Europe)	clinicaltrialsregister.eu
European Union: Health	europa.eu/european-union/topics/health_fr
International Clinical Trials Registry Platform (ICTRP)	apps.who.int/trialsearch
International Behavioral Trials Network (IBTN)	ibtnetwork.org
SCImago Journal & Country Rank: Medical and Scientific Journals Inventory	scimagojr.com
Kalya Research	www.kalya.ai
Motrial	motrial.fr
Association of Clinical Research Organizations (CRO)	acrohealth.org
Nishare	nishare.fr
Blog on NPIs	blogensante.fr/en
Non-pharmacological Intervention Researchers and Institutions world register (NIRI)	niri.fr
Plateforme CEPS	plateforme-ceps.fr
Preferred Reporting Items for Systematic Reviews and Meta-Analyses (PRISMA)	prisma-statement.org
Prospective Register of Systematic Reviews (PROSPERO)	crd.york.ac.uk/prospero
PubMed	ncbi.nlm.nih.gov/pubmed
U.S. Agency for Healthcare Research and Quality (AHRQ)	effectivehealthcare.ahrq.gov

© The Author(s), under exclusive license to Springer Nature Switzerland AG 2021
G. Ninot, *Non-Pharmacological Interventions*,
https://doi.org/10.1007/978-3-030-60971-9

Society

Academic Consortium for Integrative Medicine and Health	imconsortium.org
European Association of Psychosomatic Medicine	eapm.eu.com
European Public Health Association	eupha.org
European Society of Integrative Medicine	european-society-integrative-medicine.org
French Integrative and Complementary Medicine Society (CUMIC)	cumic.net
International College of Psychosomatic Medicine	icpmonline.org
International Society of Traditional, Complementary, & Integrative Medicine Researchers	iscmr.org
International Society of Behavioral Medicine	isbm.info
Multinational Association of Supportive Care in Cancer	mascc.org
Society for Integrative Oncology	integrativeonc.org
World Federation of Public Health Associations	wfpha.org

Practice

CAMbrella Program	cam-europe.eu/library-cam/cambrella-research-reports
Center for Integrative Medicine (University of Maryland School of Medicine, USA)	cim.umaryland.edu
Center for Integrative Medicine (Stanford University, USA)	stanfordhealthcare.org/medical-clinics/integrative-medicine-center.html
Chinese Clinical Trial Registry (ChiCTR)	chictr.org.cn
Commission of Practitioners in Alternative Medicine of Quebec (CPMDQ)	cpmdq.com
Endeavour College of Natural Health (Australia)	endeavour.edu.au
EUROCAM	cam-europe.eu
Global Wellness Institute	globalwellnessinstitute.org
Institute for Complementary and Integrative Medicine (Switzerland)	iki.usz.ch
Norway's National Research Center in Complementary and Alternative Medicine (Norway)	nafkam.no/en
Organisation for Economic Co-operation and Development (OECD)	oecd.org
School of Traditional Chinese Medicine (China)	jichu.bucm.edu.cn

Regulation

European Commission for Health	ec.europa.eu/health
European Food Safety Authority (EFSA)	efsa.europa.eu/fr
European Medicines Agency (EMA)	ema.europa.eu/ema
French National Authority for Health (HAS)	has-sante.fr
National Center for Complementary and Integrative Health (NCCIH)	nccih.nih.gov
U.S. Food and Drug Administration (FDA)	fda.gov/home
WHO Traditional, Complementary and Integrative Medicine	who.int/ traditional-complementary-integrative-medicine

Index

A

Academic Consortium for Integrative
    Medicine and Health, 248
Acceptance, 101
Acupuncture, 90, 99, 151
Acute intermittent porphyria, 176
Adapted physical activity (APA), 129
Adverse effects, 143, 157
Aerobic exercise, 124
AIR+R (post-rehabilitation network), 174
Alternative medicine, 8, 9, 14–15, 21, 23,
    241, 279
    and ancestral health practices, 145
    CAM, 147, 160
    cognitive biases (*see* Cognitive biases)
    diagnosis, 159
    ergo propter hoc fallacy, 146
    and evidence-based NPIs, 149
Alzheimer's disease, 70
Amazon, 204
American Horticultural Therapy Association
    (AHTA), 34
American ORBIT model, 256
Animal-assisted therapy (AAT), 33
Anti-aging medicines, 186, 250
Anticipation, 103
Anti-doping control agencies, 178
Anxiety, 104, 120
Art therapies, 32
Art therapy interventions, 175
Artificial intelligence, 145, 244
Artistic inspiration, 175
Assessment protocols, 235
Asterisks, 144
Augmented reality (AR), 37

Auriculotherapy, 130
Australian Research Centre for
    Complementary and Integrative
    Medicine (ARCCIM), 9
Automatic processes, 93
Ayurvedic medicine, 149
AYUSH, 10

B

Back pain
    mild/nonspecific low back pain, 120
    osteopathic method, 121
    psychotherapy, 121
    recommendations, 122
    yoga, 122
Balneological programs, 35
Balneotherapy, 89
Bariatric surgery, 72
Behavioral interventions, 55
Beta-testers, 207
Bienvieillirinm, 26
Big data analysis, 68
Bio-behavioral factors, 109
Biochemical treatments, 70, 72, 73
Biocompatible nanotechnologies, 48
Biofeedback, 93
Biological mechanism
    cellular level, 91
    molecular level, 92
    NPIs, 89
    organ and tissue level, 89, 90
Biologically active molecules, 107
Biomedical therapies, 22
Biomedical treatments, 38, 55, 81, 136, 183